The Technology Fallacy

Management on the Cutting Edge Series

Published in cooperation with *MIT Sloan Management Review*
Edited by Paul Michelman

The AI Advantage: How to Put the Artificial Intelligence Revolution to Work,
Thomas H. Davenport

The Technology Fallacy: How People Are the Real Key to Digital Transformation,
Gerald C. Kane, Anh Nguyen Phillips, Jonathan R. Copulsky, and Garth R. Andrus

The Technology Fallacy

How People Are the Real Key to Digital Transformation

Gerald C. Kane, Anh Nguyen Phillips,
Jonathan R. Copulsky, and Garth R. Andrus

The MIT Press
Cambridge, Massachusetts
London, England

© 2019 Massachusetts Institute of Technology

Some material in this book has appeared previously in *MIT Sloan Management Review* publications, including the following:

Chapter 1. Digital Disruption Is No Secret
Digital Transformation at John Hancock G. C. Kane, D. Palmer, A. N. Phillips, D. Kiron, and N. Buckley, "Coming of Age Digitally: Learning, Leadership, and Legacy," 2018 *MIT SMR*/Deloitte Report on Digital Business.

Chapter 2. Digital Disruption Is Really about People
Adobe: Combining Employee and Customer Experience G. C. Kane, D. Palmer, A. N. Phillips, D. Kiron, and N. Buckley, "Aligning the Organization for Its Digital Future," 2016 *MIT SMR* Report on Digital Business.

Chapter 3. Moving Beyond the Digital Transformation Hype
Walmart: Digital Maturity and the Long Game G. C. Kane, D. Palmer, A. N. Phillips, D. Kiron, and N. Buckley, "Aligning the Organization for Its Digital Future," 2016 *MIT SMR* Report on Digital Business.

Chapter 4. Digital Strategy for an Uncertain Future
MetLife's Digital Lens on Business Strategy G. C. Kane, D. Palmer, A. N. Phillips, D. Kiron, and N. Buckley, "Achieving Digital Maturity," 2017 *MIT SMR* Report on Digital Business.

Chapter 9. Making Your Organization a Talent Magnet
Developing a Digital Talent Strategy at Cigna G. C. Kane, D. Palmer, A. N. Phillips, D. Kiron, and N. Buckley, "Achieving Digital Maturity," 2017 *MIT SMR* Report on Digital Business.

Volvo Balances Competing Concerns for Digital Innovation Fredrik Svahn, Lars Mathiassen, Rikard Lindgren, and Gerald C. Kane, "Mastering the Digital Innovation Challenge," *MIT SMR* 58, no. 3 (Spring 2017), https://sloanreview.mit.edu/article/mastering-the-digital-innovation-challenge/.

Chapter 11. Cultivating a Digital Environment
Salesforce's Intentional Culture G. C. Kane, D. Palmer, A. N. Phillips, D. Kiron, and N. Buckley, "Aligning the Organization for its Digital Future," 2016 *MIT SMR* Report on Digital Business.

Slack—Building Culture in a Fast-Moving Environment G. C. Kane, D. Palmer, A. N. Phillips, D. Kiron, and N. Buckley, "Aligning the Organization for its Digital Future," 2016 *MIT SMR* Report on Digital Business.

Chapter 12. Organizing for Agility
Cross-Functional Hospitality at Marriott G. C. Kane, D. Palmer, A. N. Phillips, D. Kiron, and N. Buckley, "Achieving Digital Maturity," 2017 *MIT SMR* Report on Digital Business.

Chapter 13. Strength, Balance, Courage, and Common Sense
Transactive Memory at Discover Financial Services Paul Leonardi, interviewed by David Kiron, "The Unexpected Payoffs of Employee 'Eavesdropping,'" *Big Idea: Social Business* (blog), November 6, 2014, https://sloanreview.mit.edu/article/the-unexpected-payoffs-of-employee-eavesdropping/

Cardinal Health: Collaboration "Fuses" Innovation and Culture G. C. Kane, D. Palmer, A. N. Phillips, D. Kiron, and N. Buckley, "Achieving Digital Maturity," 2017 *MIT SMR* Report on Digital Business.

This book was set in Stone Serif Medium by Westchester Publishing Services. Printed and bound in the United States of America.

Library of Congress Cataloging-in-Publication Data

Names: Kane, Gerald C., author.
Title: The technology fallacy : how people are the real key to digital transformation / Gerald C. Kane, Anh Nguyen Phillips, Jonathan R. Copulsky, and Garth R. Andrus.
Description: Cambridge, MA : MIT Press, [2019] | Series: Management on the cutting edge | Includes bibliographical references and index.
Identifiers: LCCN 2018030557 | ISBN 9780262039680 (hardcover : alk. paper)
Subjects: LCSH: Information technology--Management. | Technological innovations--Management. | Organizational change. | Organizational behavior.
Classification: LCC HD30.2 .K3537 2019 | DDC 658/.05--dc23 LC record available at https://lccn.loc.gov/2018030557

10 9 8 7 6 5 4 3 2 1

Contents

Series Foreword

The world does not lack for management ideas. Thousands of researchers, practitioners, and other experts produce tens of thousands of articles, books, papers, posts, and podcasts each year. But only a scant few promise to truly move the needle on practice, and fewer still dare to reach into the future of what management will become. It is this rare breed of idea—meaningful to practice, grounded in evidence, and *built for the future*—that we seek to present in this series.

Paul Michelman
Editor in chief
MIT Sloan Management Review

Acknowledgments

As with any such endeavor, there are many people who made this book possible—far beyond the four of us. And while it's impossible to list all of the people who have inspired us and helped us get here, we want to recognize several specific individuals.

We'd like to start by thanking Doug Palmer, David Kiron, and Natasha Buckley, who have been our research partners for the past seven years, exploring the world of digital and social business. Natasha, special thanks to you for your intimate knowledge of the data and for all of your help in making the book a reality. We'd also like to thank the rest of the research team that worked with us over the years, digging through data, interview transcripts, and program schedules, and taking on tasks big and small: Al Dea, Swati Garg, Nina Kruschwitz, Saurabh Rijhwani, Dan Rimm, Negina Rood, and Allison Ryder. And thanks to our data coders: Lauren D'Alessandro, Gabrielle Hanlon, Danni Bianco, Katie Gold, Julia MacDonald, and Anna Copman.

Thank you to Paul Michelman for his support and belief in the project, and to the leaders at Deloitte who have sponsored or supported this effort: Mark Cotteleer, Rob Frazzini, Nidal Haddad, John Hagel, Alicia Hatch, Suzanne Kounkel, Andy Main, Jeff Schwartz, and Erica Volini.

A number of others contributed to the development of the book, including: Carrie Brown, Virginia Crossman, Heather Graubard, Lisa Iliff, Kelly Monahan, Stacey Philpot, Brenna Sniderman, Debbra Stolarik, and Emily Taber. And a heartfelt thanks to the dozens of executives who took time out of their schedules to talk with us over the years.

Finally, we'd like to thank all of our family and friends who have supported us through this effort. Thank you for your encouragement and for sacrificing the late nights and weekends that we would otherwise have spent with you.

Introduction: Digital Disruption—
The Cyclone Has Arrived

Andrew Grove, one of Intel's founders, warned years ago that only the paranoid survive. Paranoia seems well warranted at this particular point in time. Companies and entire sectors are being toppled at unprecedented rates. Fierce new competitors, using business models that were once inconceivable, owe their existence largely to rapidly evolving technology. If you're leading a well-established organization, you can't help but be concerned. After all, who wants to be the next Borders or Blockbuster?

This is a book about disruption caused by technology. More specifically, it's about how to manage disruption, adapt to disruption, and thrive in a world and a time marked by disruption. But it's not about technology per se. In preparing to write this volume, we reviewed many books and articles about digital disruption. A vast body of management literature describes how technology gave rise to and continues to fuel the digital disruption phenomenon. We leave guidance about technology stacks, architectures, and roadmaps to others. We have chosen to focus, instead, on the organizational changes required to harness the power of technology. For ease of discussion, we simply use the term "technology" throughout this book to refer to various digital technologies as a group.

Our focus on the people and the organizational side of digital disruption does not mean we believe that technological aspects are unimportant. On the contrary, we certainly recognize that the technological challenges facing many companies are significant. The purpose of this book, however, is simply to argue that the organizational challenges of digital disruption are on par with the technological ones, even though they have received less attention in both literature and practice. Furthermore, while the technological challenges many companies face will vary considerably by industry and strategy, our research suggests that many companies are likely to face a common set of organizational challenges.

In this book, we provide the insights necessary for leaders to navigate the journey through the strange new competitive environment wrought by digital disruption. By leaders, we mean managers at all levels in an organization who can influence how their organization works—from c-suite executives to hands-on project managers and everywhere in between. How leaders at each level adapt may differ, but adaptation at all levels is critical. C-suite executives may need to cast a bold new vision for how their organizations will adapt to a changing world. Project managers must create an operating environment that's more conducive to effective work in a digital age, innovations that can then spread across the organization as those team members experience a better way of working. Both top-down and bottom-up innovation are essential for becoming a digital organization. Executives cannot simply impose change on organizations, yet grassroots change is unlikely to be sustainable without strong executive support.

As we searched for a way to frame our research insights and guidance, we found ourselves returning repeatedly to a familiar story from our childhoods, the story of the Wizard of Oz. Most of us know the 1939 MGM movie that made a star out of sixteen-year-old Judy Garland. The film's famous lines, including "There's no place like home" and "Toto, I've a feeling we're not in Kansas anymore," are well-known staples of popular culture. The Wizard of Oz story is the story of Dorothy, a young Kansas farm girl, who, after being knocked unconscious in a cyclone that carries her house to the land of Oz, follows the yellow brick road, alongside her dog, Toto, to the Emerald City to meet the Wizard of Oz. Along the way, Dorothy encounters a cast of characters, including the Scarecrow, the Tin Man, the Cowardly Lion, and the Wicked Witch of the West.

The movie starts with the cyclone, which seems like an apt metaphor for digital disruption. Dorothy didn't choose to go to Oz but was swept up by forces beyond her control. Her world in Kansas is shown in black and white, in stark contrast to the Technicolor land of Oz, where her house lands. Dorothy has no choice but to navigate this new landscape, developing new friendships and facing unfamiliar challenges, to find her way home. Aspects of the cyclone accurately characterize how many companies experience digital disruption. For most, it's a journey that companies engage in because they have no choice. They have been swept up by forces beyond their control, taken to a new world where the rules of competition seem as different as the contrast between the black-and-white hues of Kansas and the Technicolor palette of Oz.

Perhaps the most important observation about the cyclone, however, is that the story of the Wizard of Oz isn't really about the cyclone. Dorothy's adventures certainly would never have happened had the cyclone not come to Kansas, but the story is more about Dorothy making her way in this strange new world than it is about how she got there in the first place. In the same way, the story of digital disruption we explore here isn't really about technology. Rather, it is about how companies navigate their way through the new competitive environment to which technology has brought us. It is about learning to do business in different ways, restructuring organizations to enable them to respond more effectively to changes brought by an increasingly digital environment, and learning to adapt individual and institutional skill development and leadership style for the demands of this rapidly changing world. This cyclone of digital disruption didn't just touch down recently. It has a long path tracking across industries and professions for decades. Although we don't know exactly the details of the next stage of digital disruption, we have no reason to believe that its impact will abate. While exploring the conditions that got us here may be instructive, the story that most readers are concerned with is what to do once we find ourselves in this strange new world of digital disruption.

The Wizard of Oz movie is based on a series of books by L. Frank Baum, beginning with *The Wonderful Wizard of Oz*, published in 1900. A critical difference between the movie and the book is that in the book, Dorothy realizes that she can never return permanently to Kansas and remains in Oz. Like the heroine of the book, companies can never return to the predigital disruption world they once inhabited. We hope that you find the metaphor useful in understanding the digital disruption challenges and opportunities organizations face. As Dorothy would say, "I've a feeling we're not in Kansas anymore."

About the Research

This book is grounded in four years of research, conducted in partnership with *MIT Sloan Management Review* and Deloitte, into how technology changes the way companies operate. We surveyed more than 16,000 people over four years about their experience with digital disruption and their perceptions of the nature and adequacy of their organization's response. Although some questions remained consistent from year to year, we also asked new questions that arose from the results of the previous year's research. Each year's survey consists of between 3,700 and 4,800 responses. When discussing survey results in the book, we typically refer only to results from a single year, but we do not specify from which year the data are drawn for ease of reading. While there may be some small shifts in the actual numbers in the data from year to year, the general relationships and the key themes and takeaways have remained consistent over the four years we collected these data. If readers would like to peruse the published reports, they are all available online at the *MIT Sloan Management Review* website.[1]

The survey approach has multiple advantages. The method is familiar to the most people, can be relatively easily implemented and analyzed, can be widely disseminated, and can improve through iterations. As such, it's a good approach to studying a rapidly evolving but unevenly distributed phenomenon. Nevertheless, survey responses must be utilized carefully since respondents' perceptions are often limited and biased. In some situations, these perceptual data are not an issue. For example, when we asked employees how long they plan to stay in their current jobs in light of digital trends, we believe the responses reliably reflected their intentions. In other cases, respondents' perceptions were more problematic. For example, when we asked respondents to evaluate the digital maturity of their organization, their perceptions may not match objective reality, since they have a limited ability to calibrate their organization to others. We highlight those limitations explicitly when we believe it is important to do so, but we make a blanket disclaimer here to call attention to the inherent limitations of survey data that persist throughout this book. Our language in every chapter reflects this cautionary use.

We also conducted more sophisticated statistical analyses on our data to ensure our findings hold up under more intense scrutiny. We performed multiple regression analyses to confirm that our findings are not purely the result of some other company characteristic, such as company size or age. Additional analyses ensured that our findings are not the result of other factors, such as the specific survey instrument we used. For example, factor analysis demonstrated that respondents weren't simply rating their company high or low across the board rather than in response to specific questions—a characteristic known as *common method bias* by researchers. While the results we present in this book hold up to more stringent statistical scrutiny, we present them in the simplest possible terms for ease of presentation and discussion.

We also use two other types of data to augment our survey findings. First, we interviewed more than seventy-five thought leaders at companies like Walmart, Google, MetLife, Salesforce, Marriott, and Facebook, just to name a few. Interviews enable us to explore in greater depth how technology is changing companies, and our interviewees' experiences help us interpret the data from our surveys. We reveal the identities for some, but others asked us to mask their identities. These interviews provide context to our survey findings, help validate the insights we derive from them, and provide additional perspectives from people on the front lines of digital disruption.

Second, we augment the insights of our primary research by drawing on established literature in the fields of information systems, management, marketing, psychology, and operations to set our findings in the broader context of management science. By triangulating our data using multiple and dissimilar sources—quantitative and qualitative methods, primary and secondary data—we hope to offset the limita-tion of any single data source to provide a balanced, authoritative, and novel insight into the problem of how companies respond to digital disruption.

We also don't want to be bound entirely by the data. Doing so would leave us with a backward-looking approach, no matter how timely, that simply describes how companies have responded. Instead, we use this data to make proactive and prescriptive recommendations on how managers should respond. We hope to balance our insights with differ-ent types of experience to provide the most holistic picture. Our author team consists of both academics researching and teaching about digital disruption for decades, and consultants who have been actively work-ing with companies to help them adapt to the challenges of digital disruption for far longer.

Throughout the book, we tend to refer to our collective experience as "we," even if it represents only a portion of the four-person author team. Each member of the team has contributed different expertise and experiences. For example, we rely extensively on Kane's academic research, conducted with various coauthors, to inform our findings. Likewise, Andrus is currently working actively with clients on these issues. Phillips led this research project from the Deloitte side from the beginning, so she is deeply familiar with the data and findings. Copul-sky, a recently retired Deloitte principal, a widely published author, and a current faculty member at Northwestern University, sponsored this research and brings a talent for distilling complex ideas into compel-ling narratives. Our team worked together remarkably well in the devel-opment of this book, and we believe the whole is more than the sum of the individual contributions. Despite the individual roles played by team members, we speak with a singular voice to reflect this integrated approach to the project.

By bringing together different experiential perspectives to our data, we hope to offer a clear voice, supported by research, for how companies should respond to digital disruption. In doing so, we hope to navigate between what we regard as the limitations of previous books on this theme—avoiding either an overly analytic approach that is divorced from what most companies are dealing with today or an overly shallow, hype-driven treatment that delivers little substance with which readers can actually do anything concrete. We hope the result is a balanced, authoritative treatment that provides leaders with actionable guidance about how to lead their organization into the future at whatever level their influence allows.

What to Expect

We divide this book into three different parts. In the first part, we deal with the phenomenon of digital disruption and how companies should respond. In this part, we argue that companies should think of themselves as adapting to a changing environment.

- Chapter 1 posits that most leaders and employees know digital disruption is happening, yet they are not acting according to that knowledge. In this book, we focus on how to respond to digital disruption (vs. debating whether it's occurring).

- In chapter 2, we position digital disruption as being primarily about people, specifically the different rates at which individuals, organizations, and societies respond. Any efforts for organizational digital transformation, therefore, must also involve changes to how people work.

- In chapter 3, we define the concept of *digital* as an adjective (not a noun) and introduce the key construct of our book—digital maturity. We argue that digital maturity is the goal to which most companies should aspire when trying to compete.

- Chapter 4 addresses the importance of business strategy for successfully navigating a competitive environment increasingly infused with technology. We introduce approaches for developing strategy in a rapidly changing environment.

- Chapter 5 uses the metaphor of duct tape to introduce and explain the academic concept of affordances. We discuss how affordances—a focus on what technology allows companies to do differently—is a valuable way of thinking about the strategic challenges companies face.

In the second part, we deal with the implications of digital maturity on leadership, talent, and the future of work.

- Chapter 6 introduces the characteristics of digital leadership. Many people tend to believe that leadership is fundamentally different in a digital environment. We suggest that the basics of leadership are the same, just enacted in a different environment.

- In Chapter 7, we go beyond the universal basics of good management introduced in chapter 6 to address the specific skills and capabilities unique to the digital environment. We describe what skills and capabilities are necessary for leading a digital organization and which of these are most lacking.

- Chapter 8 introduces the digital talent mindset. We assert that the key way employees need to respond to digital disruption is through continual learning. Employees want to develop digital skills, but organizations are not supporting their efforts.

- Chapter 9 shows that most companies feel like they are lacking sufficient talent to compete in a digital world. We provide strategies that organizations can follow to attract and retain these types of valuable employees.

- Chapter 10 focuses on the future of work. We examine what work will be like in the coming decade as a result of digital disruption.

In the third part, we address the conditions for successfully adapting to digital disruption that most organizations will need to create. Skip ahead if you'd like, but don't quit reading before you get to this part!

- Chapter 11 deals with cultivating a digital environment at your company. We find that several organizational characteristics are associated with digital maturity, namely, being exploratory, distributed, collaborative, agile, and data driven. We deal with these issues in greater depth in this section.

- In Chapter 12, we focus on how digitally maturing companies are organized differently. We find that they are far more likely to rely on cross-functional teams and push decision making to lower levels of the organization. This organizational structure helps the organization respond to changes more quickly.

- Chapter 13 addresses how digitally maturing companies are more intentionally collaborative. We explore the benefits of technology to enable stronger collaboration, but also how these tools must be used intentionally, lest they actually lead to inferior outcomes.

- Chapter 14 focuses on how companies can cultivate a more experimental mindset. In a changing digital environment, companies should experiment and iterate, but these values are directly in contrast with many of those that have defined companies for most of the last fifty years.

- In chapter 15, we provide practical guidance to begin making your company more digitally mature. We suggest an approach for measuring digital maturity in your organization and for moving forward toward maturity.

I Navigating Digital Disruption

1 Digital Disruption Is No Secret

We set aside our temptation to open this chapter with a compelling case for why organizations need to urgently adapt to the disruptions wrought by technology. Why? Because, as Canadian singer-songwriter Leonard Cohen writes, "Everybody knows. Everybody knows that's the way it goes. Everybody knows."

You may have witnessed firsthand the destruction of the newspaper, recorded music, and photographic film industries during the first wave of technology that began nearly two decades ago. You likely know that similar disruptions are currently under way in the hotel, taxi, and retail industries, as the digital cyclone continues to reshape the business landscape through companies like Uber, Airbnb, and the omni-business behemoth Amazon. You also know that digital disruption is far from over, as analytics and data science come into their prime and other technologies, such as artificial intelligence, blockchain, virtual and augmented reality, and autonomous vehicles, loom on the horizon. You know that more businesses, and entire sectors, will certainly topple, with new, nimble competitors scooping up current customers and bringing new customers into the fold.

Our survey confirms our assertion that everybody knows that digital disruption is real and happening now. When we asked survey respondents, "To what extent do you believe that digital technologies will disrupt your industry?" 87 percent replied that these technologies would disrupt their industries to a great or moderate extent. Only 3 percent of respondents believed that technology is likely to have no effect on their industry (we didn't ask this 3 percent when was the last time they had read a hardcopy newspaper, visited a record store, or booked a trip through a personal travel agent). Most respondents also recognized the need for organizations to adapt effectively to these changes as an important factor for success and survival in the coming years—84 percent of our respondents agreed or strongly agreed that becoming a digital business is important to their organization's success.

So, let us stipulate from the outset that digital disruption is happening and will likely affect your industry. Precisely when and how that disruption will happen may differ, but most of us can agree that it's happening in some way and to some level. Such stipulation allows us to turn our attention to where it really matters—what to do now that it is happening.

The Gap between Knowledge and Action

Knowing digital disruption is happening and doing something about it are entirely different matters. The philosopher William James noted, "Thinking is for doing," meaning that the purpose of knowing things is to act in accordance with that knowledge. You might expect, consequently, that every organization has a well-developed strategy and action plan for responding to these disruptions. But the reality is quite different, just as homeowners in hurricane- or cyclone-prone areas often seem caught off guard when the actual storm strikes. We asked survey respondents whether their companies were adequately preparing for the digital disruptions most likely to occur in their industries (figure 1.1): 44 percent said their companies are doing enough; 31 percent said their companies aren't doing enough; and 25 percent neither agreed nor disagreed that their companies are doing enough. The gap between the 87 percent who said digital disruption will affect their industry and the 44 percent who said their company is doing enough is, in a word, staggering. Everybody (well, almost everybody) knows digital disruption is happening. Yet, only a minority report that their organizations are doing enough to respond effectively.

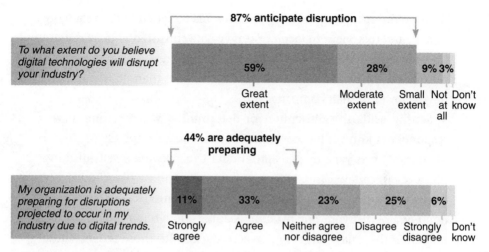

Figure 1.1

Interpreting Survey Data

These survey responses exhibit many nuances that merit discussion. Perhaps our average survey respondent does not actually know how much response to this disruption is "adequate." Given the uncertainty regarding the rate and scope of change from digital disruption, predicting how much an industry is disrupted, when that disruption will happen, and what will be the appropriate response can be extremely difficult even for industry experts—let alone the average employee who responded to our survey.

Yet, this data should not be dismissed entirely simply because of this limitation. In many ways, employees may actually be *better* suited to know if an organization is responding adequately, because they have firsthand experience about how these responses play out in interactions with customers, suppliers, and colleagues. As we assemble these disparate responses into a coherent dataset, employees of various ages, with different backgrounds and skill sets, may have a more complete understanding of what may be possible with technology and could actually provide a better idea of what constitutes an adequate response than would the average c-suite executive. Executives often have a more rarified and rosy view. In fact, our survey data consistently demonstrate that c-level and board-level respondents tend to report a far more optimistic view of their company on important issues than do respondents ranked lower in their company.

Setting aside the difficulties of determining whether our survey respondents know what constitutes an adequate response to digital disruption, this type of perceptual data can become a self-fulfilling prophecy in certain cases. Whether employees know what an adequate response to digital disruption is may not really matter. If your employees don't *believe* that your organization is doing enough to respond to digital trends, they may start acting in accordance with those beliefs. Those employees with desirable skills may leave their jobs to join companies they believe are responding adequately. In fact, this trend is occurring, a topic we deal with in greater depth in chapter 9. Perceptions matter because employees act on their perceptions. If employees perceive that their employers are not doing enough to respond to digital disruption, they may leave, and companies may find themselves with insufficient talent to respond adequately when the time comes. Whether your organization's leaders *believe* they are doing enough to respond to digital disruption may be immaterial. It may even be immaterial that the organization is, in fact, doing enough. What your employees think is happening and whether they believe it is adequate may matter most.

This insight provides an important takeaway for managers. Because we all know that digital disruption is happening, an effective organizational response to that disruption needs to involve both talk *and* action. If companies simply talk about being a digital company but don't back up that talk with concrete behaviors, employees will notice the discrepancy and respond accordingly. Likewise, an organization taking active steps to respond to digital disruption in some secret or remote innovation lab, without communicating those initiatives to employees or involving them, may also be problematic. An effective response to digital disruption must involve all aspects of the organization and both concrete action and clear communication regarding how the organization is changing.

Why Aren't Companies Acting?

Why aren't companies responding with greater urgency to the threat of digital disruption? Executives may not understand enough about technology to make the changes or to understand the urgency necessary. Board members and investors may care more about short-term profits than the long-term viability of the company. Many leaders may be just counting down the years to retirement, and thus they don't have the energy or the interest to engage in the types of changes necessary to adapt the company for a future in which they will not participate. While any of these scenarios may be true, the most common reason we encounter is that companies are simply trying to balance too many competing priorities. It's difficult to keep the current business running while also preparing it for a digital future.

Our data rule out one possible explanation for companies' lack of responsiveness. We wondered whether some executives just didn't think digital business was important for their organization's or industry's future. Might digital disruption affect the industry but not a particular company in it? When we asked whether being a digital business was important to the success of respondents' organization, a whopping 85 percent either agreed or strongly agreed that it was. Even though their organizations still weren't doing enough to respond, the vast majority of respondents recognized that digital business was critical to their future.

The gap between knowing about a need and taking action on that need is a well-documented phenomenon not confined to digital disruption. Jeffrey Pfeffer and Robert I. Sutton described the "knowing-doing gap" in a Harvard Business School Press book in 1999. As they write in their book's first chapter, "Why do so much education and training, management consulting, and business research and so many books and articles produce so little change in what managers and organizations actually do?" They assert that the first step in resolving the gap is focusing on "why before how." They argue that "too many managers want to learn 'how' in terms of detailed practices and behaviors and techniques, rather than 'why' in terms of philosophy and general guidance for action."[1]

Why respond to digital disruption? Put simply, the emergence of new classes of technologies, such as social media, mobile technologies, big data analytics, artificial intelligence, blockchain, additive manufacturing, autonomous vehicles, and augmented and virtual reality, change what is possible for business. Leaders who want to maintain advantage, discover new opportunities, and better serve their customers will leverage the opportunities provided by these technologies to do business differently. Yet, leveraging these technologies amid a rapidly evolving digital infrastructure requires some fundamental shifts in how companies are organized. This book provides the tools to meet these goals.

Competency Traps: What Got You Here Won't Get You There

Established companies in particular typically face significant challenges from digital transformation—one of the biggest is their past success.[2] This is often referred to in the management literature as a *competency trap*. Competency traps are beliefs that the factors of past success will also lead to future success. Technology is changing the competitive landscape—providing new ways of delivering value to customers and new service opportunities—and past success factors may not be associated with future success. If companies don't change their processes and mindsets to take advantage of new opportunities for doing business made possible through emerging digital infrastructure, then established or new competitors likely will.

GE is one example of a company seeking to overcome competency traps. In the 1990s GE became known for its adherence to a process known as Six Sigma, a set of techniques for reducing the error rate in manufacturing processes to 0.00033 percent. Six Sigma was a key factor associated with GE's success during the 1990s and early 2000s. Six Sigma, however, also has its limitations. It tends to be exceedingly difficult—if not impossible—to maintain Six Sigma standards while also experimenting with new ways of doing business. The process is not conducive to the types of agile responses to environmental shifts that characterize the world of digital business.

To address the need for faster change and agility, GE has developed a complementary approach known as FastWorks, which combines Lean Startup principles (created by Eric Ries) with GE's size and resources. But, as GE's culture transformation leader Janice Semper remarks, "It's been harder in some of our businesses where they don't have as much volatility or disruption yet. They don't necessarily feel the same need as those parts of our businesses that are being disrupted or are in extremely challenging, volatile, ambiguous environments[,] where this is really the only way to go forward and work." Semper described two other "interventions" introduced to accelerate changes to the legacy company. First, GE leaders introduced new company values, which are "customers determine our success; stay lean to go fast; learn and adapt to win; empower and inspire each other; and deliver results in an uncertain world." Second, they redesigned the performance-management system. "We moved away from our annual, traditional performance-management process to a more continuous and fluid system that is in sync with the FastWorks way of working and with the GE Beliefs mindset."

"We Have Met the Enemy and He Is Us"

Another key reason for the disparity between awareness of and action on digital disruption is that many executives simply don't understand how quickly this threat may emerge. Many are waiting to see evidence for this disruption in their organization's bottom line before acting. By the time that evidence appears, however, it may be too late. Lagging indicators are not effective tools for identifying and assessing looming threats. For example, profits in the newspaper industry grew steadily well into the dot-com boom, at which point the profits fell off a cliff. Executives need early warning systems, particularly since many may overestimate their ability to respond in time, thinking they can just run out and invest heavily in technology when the threats manifest. Many of the leaders we

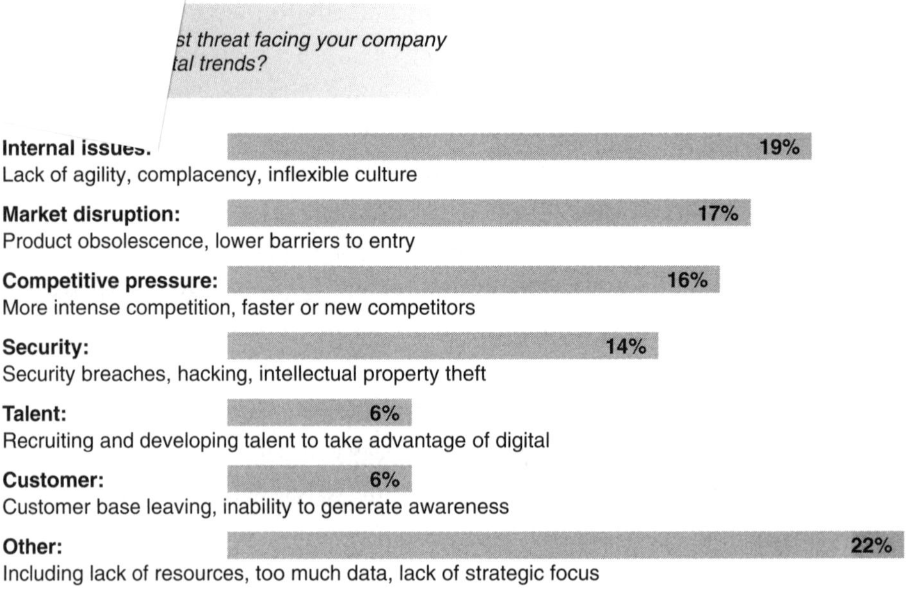

st threat facing your company
tal trends?

Internal issues. Lack of agility, complacency, inflexible culture	19%
Market disruption: Product obsolescence, lower barriers to entry	17%
Competitive pressure: More intense competition, faster or new competitors	16%
Security: Security breaches, hacking, intellectual property theft	14%
Talent: Recruiting and developing talent to take advantage of digital	6%
Customer: Customer base leaving, inability to generate awareness	6%
Other: Including lack of resources, too much data, lack of strategic focus	22%

Figure 1.2

interviewed from companies putting the most effort into digital transformation today are privately wondering if they may have already waited too long. We have seen a significant increase in the number of legacy companies beginning to place significant emphasis on their digital efforts.

Although only a minority of respondents said that their organizations viewed digital trends as a threat as opposed to an opportunity, we asked them about the specific nature of the threat their companies faced as a result of digital trends. Instead of providing a selection of response choices, we asked them to use their own words in an open-text response. Our research team then coded all 3,300 free responses into larger categories (figure 1.2).

The biggest threat that respondents reported falls under internal organizational issues, such as complacency, inflexible culture, and lack of agility. In other words, the biggest threat of digital disruption is in the organization itself—that the company would be either unable or unwilling to change fast enough to respond to the threats posed by digital disruption. The cartoonist Walt Kelly created a memorable character,

Pogo, a lovable swamp-dwelling possum. In 1970, in the early days of the environmental movement, Kelly inked a cartoon for the first-ever Earth Day, featuring Pogo and a fellow swamp dweller surveying a litter-filled swamp. Pogo's observation could easily apply to the sentiments of our respondents: "We have met the enemy and he is us."[3]

The next biggest threats identified are market disruption, such as product obsolescence or lower barriers to entry, followed by increased competitive pressure from either new or established competitors. Taken together, these responses suggest a consistent story of digital disruption increasing the competitive pressure on organizations and a concern that those organizations will be unable to act quickly or adequately enough to respond to those competitive changes. To us, these sound like genuine threats to which organizations should be prepared to respond.

How Is Digital Business Different?

We asked respondents to our survey how digital business differed from more traditional ways of doing business. This was another open-text response, where we gave respondents a blank text box to respond to the question in whatever way they wanted to. Members of our team then read through all 2,362 of these responses, categorizing responses into groups of similar responses. The results can be seen in figure 1.3.

The biggest difference respondents said is the pace of doing business. Put simply, digital business requires companies to act and respond faster than they ever have before. The challenge is that many of the organization's communication and decision-making structures can't move as quickly as the organization needs them to. An executive we interviewed says, "It's the speed at which the landscape is changing through digital, allowing new competitors to play, that makes it really transformative. Not only is technology making industries work faster and more efficiently—we all have access to the same technology. It's the entrance of new competitors that we would never have thought of before that's throwing everyone a huge curveball." The pace of change associated with digital business is a key theme that pervades this book.

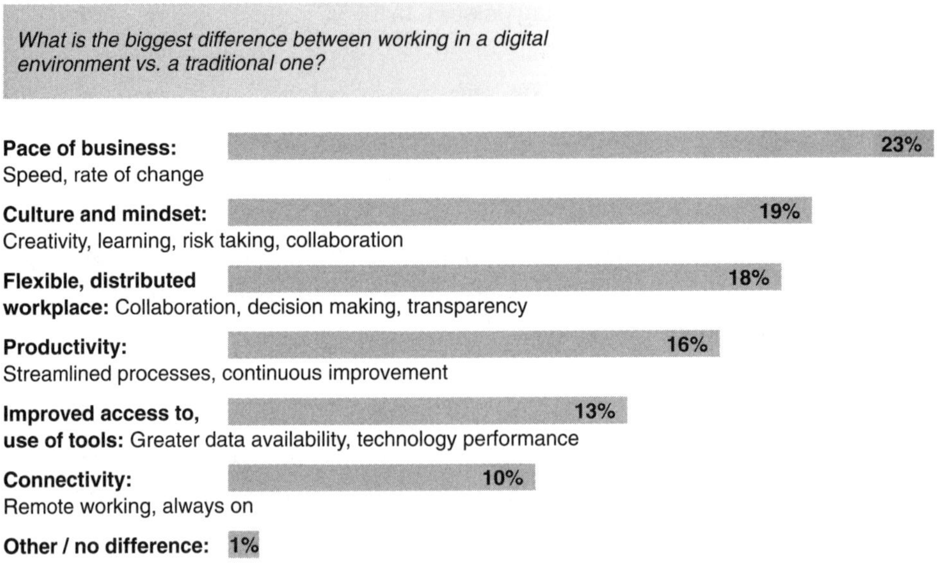

What is the biggest difference between working in a digital environment vs. a traditional one?

Pace of business: 23%
Speed, rate of change

Culture and mindset: 19%
Creativity, learning, risk taking, collaboration

Flexible, distributed   18%
workplace: Collaboration, decision making, transparency

Productivity: 16%
Streamlined processes, continuous improvement

Improved access to,   13%
use of tools: Greater data availability, technology performance

Connectivity: 10%
Remote working, always on

Other / no difference: 1%

Figure 1.3

The second most common difference is mindset and culture. These responses centered on the need for changes to organizational culture, but descriptions of these changes were not entirely positive. Respondents said that these cultural shifts create tensions with employees who have a more traditional mindset. In other words, competency traps may exist at the individual level, especially for established companies. Employees who have had success with a particular way of working in the past may be reluctant to change those ways of working for the future. We deal with culture in chapter 11.

The third most common difference respondents describe relates to organizational structure: the need for a flexible, distributed workplace. Part of this is about collaboration, how decisions are made, and how teams are organized. But this is also about rethinking teams and talent. Melissa Valentine of Stanford says, "It seems pretty clear that the boundary of the firm is changing in significant ways. I hear the 'core-periphery' idea a lot here in Silicon Valley." In that model, a company relies on a group of core employees, which the company plans to invest in and nurture while tactically leveraging networks of external on-demand talent. Valentine notes that even large companies may consist of a "core team, and then peripheral employees and projects around it," instead of full-time employees working for a single company. For some organizations, this model may require a new perspective on how to blend full-time employees with talent sourced from the open market. Recent studies find that employers expect to dramatically increase their dependence on contract, freelance, and gig workers over the next few years.[4] We deal with this topic in greater depth in chapter 12.

Finally, the fourth most cited difference was productivity, which can be a double-edged sword. According to John Hagel, co-chair of Deloitte's Center for the Edge, "If you're truly going to accelerate performance improvement, you have to stop focusing on efficiency. If it's just efficiency, that's a diminishing-returns game. The more cost effective and faster you are, the harder it's going to be to get to that next level of efficiency. But if you focus on effectiveness, on impact, on value delivered to whatever the arena is—the sky's the limit. That's a mindset shift, getting out of that efficiency mindset."

Disruption Is Here, It's Just Unevenly Distributed

The science-fiction writer William Gibson is quoted as having noted, "The future is here, it's just unevenly distributed." Likewise, while we all know digital disruption is happening, the current state is unevenly distributed across industries. Figure 1.4 shows the results of key responses to our survey, grouped by industry. We deal with the specific import of these questions later in the book, but at this point, the pattern of responses is helpful. We rank industries by answers to several questions that we find are key correlates with digital maturity.

The results show that many of the "usual suspects" are clear leaders when it comes to digital maturity—technology, telecommunications, and media. On the other hand, we were somewhat surprised to discover that no industry emerges as a clear loser. No industry ranks in the bottom five across all questions. The construction and real estate sector,

Industries born of technology lead the list of sectors with the greatest penetration of digitally maturing organizations—IT, telecom, and media & entertainment. However, this year's digital business study did not find a consistent set of laggards on the opposite end of the spectrum. Companies in each sector have strengths to build on as well as weaknesses to address.

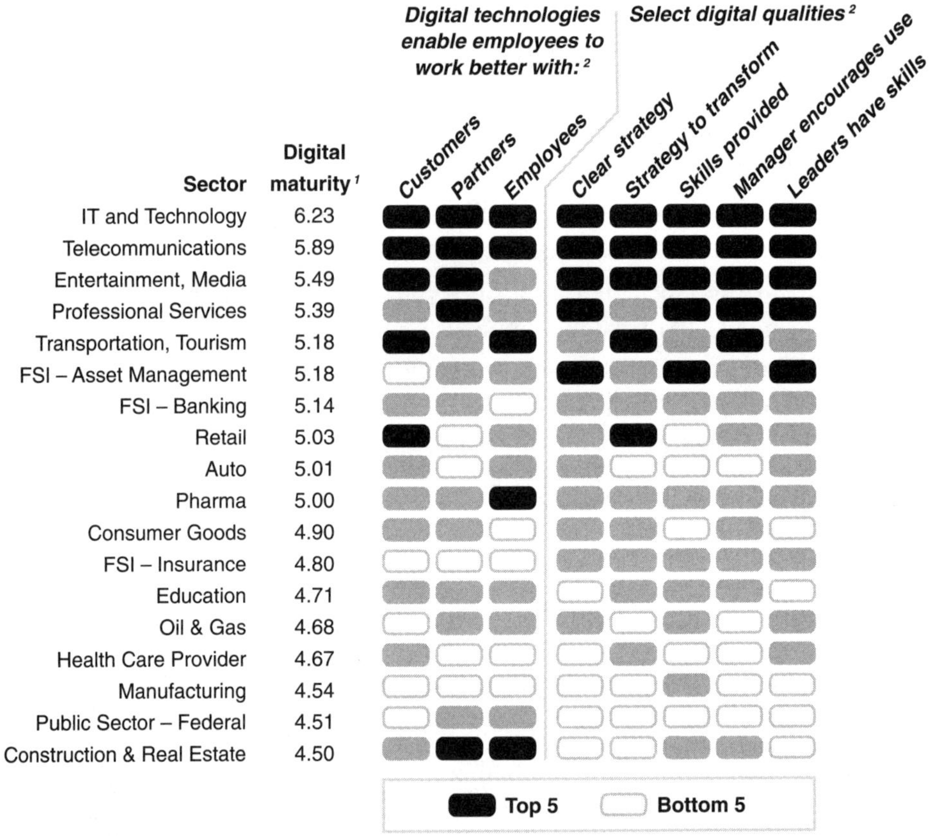

Digital technologies enable employees to work better with:[2] Customers, Partners, Employees

Select digital qualities[2]: Clear strategy, Strategy to transform, Skills provided, Manager encourages use, Leaders have skills

Sector	Digital maturity[1]
IT and Technology	6.23
Telecommunications	5.89
Entertainment, Media	5.49
Professional Services	5.39
Transportation, Tourism	5.18
FSI – Asset Management	5.18
FSI – Banking	5.14
Retail	5.03
Auto	5.01
Pharma	5.00
Consumer Goods	4.90
FSI – Insurance	4.80
Education	4.71
Oil & Gas	4.68
Health Care Provider	4.67
Manufacturing	4.54
Public Sector – Federal	4.51
Construction & Real Estate	4.50

■ Top 5   ▢ Bottom 5

[1]*Digital maturity is calculated as the average maturity of responses from a given sector. Respondents were asked to rate their organization's digital maturity on a 10-point scale with 1 being least mature and 10 being most mature.*

[2]*Correspond to specific survey questions in the study. Percentage of respondents who agree/strongly agree their organization has the relevant digital skills or capabilities.*

Figure 1.4

for example, ranks lowest in terms of digital maturity—defined as an organization where digital has transformed processes, talent engagement, and business models—but ranks in the top five industries reaping digital gains by improving work with partners and employees. Companies in this sector also lag in the development of digital strategies focused on transforming their businesses. Consumer goods companies sit squarely in the middle of the digital maturity spectrum but fall short on digitally enabling employees.

The lesson here is that while certain industries are clear leaders and can serve as aspirational models for other industries, no industry is without hope. Each likely has some strengths to build on to begin to take their companies into a digital future.

You're Never Too Old to Adapt—Individually or Organizationally

If learning and adapting is at the heart of tackling digital disruption, what does that mean for older companies (and individuals) who have been doing things the same way for years? The old adage "You can't teach an old dog new tricks" refers to the widespread perception that as people (and organizations) age, they tend to be more fixed and set in their ways. Indeed, we are amazed that children pick up languages so easily, while otherwise intelligent adults struggle.

Look at startups, often seen as digital role models. The image of the startup is an organization that is more agile and learns and adapts more quickly than established organizations. They are also seen as being more innovative and creative. Like children, their organizational brains are more flexible and therefore have tremendous ability to learn. But over time, organizations tend to produce the equivalent of the chemical inhibitors that keep adults from learning. Instead, they tend to focus on what they already know (and what has led to their history of success) to take action and to get things done. They focus on productivity and efficiency, rather than learning, growth, and innovation. The key for older, legacy organizations is to identify and work around the organizational inhibitors that keep them from learning and to foster a culture of learning and a growth mindset in the organization—something we address in part II.

Digital Transformation at John Hancock

A good example of an established company beginning to make the necessary changes to adapt to a digital world is the financial services firm John Hancock. The company's leadership realized that its business and organization needed to be refreshed for the twenty-first century and better compete in a digital world.

Barbara Goose, senior VP and chief marketing officer, wanted the company to become more innovative, more entrepreneurial, faster moving, more empowered, and more collaborative. To do so, she first needed to give the people tasked with the change the freedom to work outside the traditional bureaucratic structures that define many legacy companies. Innovation teams needed to be isolated and protected, "to be freed from the corporate shackles in some ways to be able to innovate and progress faster," said Goose.

Digital transformation can't just be a top-down mandate to change. Instead, it involves creating the conditions in which existing employees start thinking and working differently, driving change from the bottom up as well. VP of digital strategy Lindsay Sutton notes, "At the end of the day, so much is about talent. Talent is two-pronged, by skillset and by attitude. . . . [T]hat's what you need to drive an organization forward into an era they are primed to be a part of. Attitude is the one thing we sometimes forget." With the right attitude, people can develop the skills they need to work and learn in the fast-moving and ambiguous conditions that are at the heart of digital business. "People with that mentality are everywhere. Some of them need to just be reminded that they can be that person."

The executives at John Hancock see the competitive landscape shifting, are trying to create new competencies for competing in a digital world, and are trying to reshape how their employees work, think, and learn. Yet, driving the need for change can be difficult when established companies are performing well. Goose notes, "It's hard to drive change when people feel that the company has been successful doing everything the way it always has." She continues, "In looking toward the future, they can see that the world and customer needs are changing. We need to evolve and experience a revolution to get to a very different place as fast as we can, but it's hard to do that quickly in such a big company."

Takeaways for Chapter 1

What We Know	What You Can Do about It

• Everyone recognizes that digital disruption is happening, but most companies are not doing enough in response.
  ◦ 87 percent of respondents indicate that their industry is likely to be disrupted by digital trends,
  ◦ but only 44 percent reported their organization is doing "enough" about it.
• A key reason for this knowing-doing gap is the tendency of many organizations to underestimate the threat posed by digital disruption and the need to respond quickly.
	• Ask yourself if your organization is part of the 43 percentage gap between those who say digital technologies are likely to disrupt their industry and those whose companies are adequately preparing for it.
	• Start with an inventory of the opportunities and threats that technology poses for your organization.
	• Sequence the inventory based on potential impact and immediacy.
	• For each opportunity and threat, describe how you are responding.
	• Rate each response in terms of its likely effectiveness.
	• For the areas where your response is less effective than you might like, assess what you might do to enhance or improve your response.
	• Identify what's getting in the way of a more effective response and what you might need to do differently (i.e., potential interventions).
	• Finally, create an action plan for the three most urgent interventions.

2 Digital Disruption Is Really about People

Besides underestimating the threat, another reason companies may not be acting quickly enough in the face of digital disruption is that most executives don't understand exactly the key challenge facing organizations. If they don't understand the underlying nature of the problem, they can't know whether and how to respond to it effectively. In the words of Pfeffer and Sutton, they can't formulate the answer to the question of "why" the organization needs to change. Fortunately, this is a problem that our book can help address. Many treatments of digital disruption regard the rapid pace of technological innovation as the key problem facing organizations. Indeed, technological innovation is happening at a faster rate than ever before. Computers continue to become smaller, cheaper, more powerful, better connected, and embedded everywhere. While the increasing rate of technological innovation is a significant part of the challenge facing companies, it is not the problem in and of itself.

The true challenge of digital disruption facing organizations (and, indeed, a major part of the solution, as we will see) is people—specifically the different rates at which people, organizations, and policy respond to technological advances (figure 2.1). Technology changes faster than individuals can adopt it (the adoption gap); individuals adapt more quickly to that change than organizations can (the adaptation gap); and organizations adjust more quickly than legal and societal institutions can (the assimilation gap). Each of these gaps poses a different challenge for companies.[1]

Just how fast these curves are changing is under debate. For example, while technology is certainly increasing in power (and decreasing in cost)

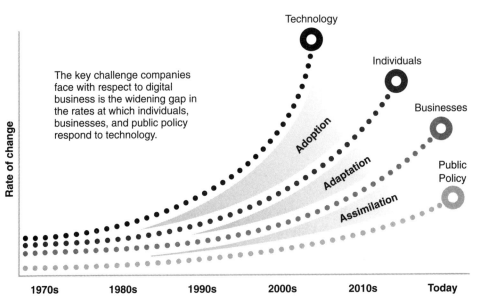

Figure 2.1

exponentially, the rate of those changes differs depending on whether you are talking about processing power (18 months, Moore's law), storage (12 months), or networking speeds (9 months). Yet, cheaper processors, more robust storage, and faster networking don't threaten organizations. The threat comes when someone realizes that this faster, cheaper, better computing environment presents new ways of solving business problems. These changes in technology use occur more unpredictably. The key point is simply that changes occur at different rates, with organizations adapting more slowly than either technology or individuals' use of it; and the gaps between these different levels of technology use are getting wider.

Adoption

Adoption describes the gap between the rate at which technology changes and the rate at which individuals make those changes a part of their daily lives. In the seminal work of Everett Rogers,[2] he labeled innovation adopters based on various rates and phases: innovators, early adopters, early majority, late majority, and laggards. The result is a cumulative adoption function by which innovation occurs rapidly as the early and late majorities begin to adopt. This adoption curve is certainly still relevant to technology companies and IT functions that are trying to drive the adoption of certain types of technology by employees in the marketplace or across the enterprise.[3]

Despite occurring to varying degrees, with a significant portion of individuals lagging behind, adoption is not the most critical digital disruption problem most managers face. Individuals generally still adopt technology faster than organizations can adapt to it. As individuals now have easy access to robust consumer-facing technology products (versus depending on their employers for these formerly costly devices and services), they are able to gain quicker fluency with new technologies. For example, Facebook is an extremely robust collaboration platform freely available to consumers, while enterprise versions of similar tools can cost upward of fifteen dollars per user, per month. Furthermore, consumer technology is increasingly user friendly and intuitive.

This situation is relatively new. As recently as ten to fifteen years ago, businesses adapted more quickly to technology than individuals did. The reason was simple economics. Prior to the first decade of this century, most people could only afford technology through their employers, and the so-called enterprise-grade technology was far more advanced than consumer-facing technology. As the costs of information technology dropped, powerful consumer-facing online platforms became widely available, and powerful mobile devices have become ubiquitous.

The rapid rise of consumer-facing platforms such as Google, Facebook, and Amazon—all currently among the top five most valuable companies in the world—are a testament to how quickly individuals have adapted to change.[4] As platforms and devices gather more data from increasing levels of user interactions, they evolve in ways that speed up the adoption curve. Facebook and other platforms engage in A-B testing to optimize every facet of platform design to make it more usable.[5] Mark Zuckerberg says that Facebook runs approximately ten thousand different versions of the platform looking for small ways to improve the experience and increase the time users spend on it.[6] As more people use the platforms, the companies gather more data about how the platforms can be modified to deepen adoption. Adoption will continue to be an issue for companies using expensive technologies that only organizations can afford, but for most technologies, the problem lies elsewhere. Most organizations don't need to drive further adoption; they need to adapt to the facility individuals have already built with these tools. The "bring your own device" (BYOD) policies implemented by many enterprise technology departments highlight and, perhaps, exacerbate the lag between adoption and adaptation.

Understanding the adoption curve isn't entirely without value, however, because it can provide important foresight for when and how certain strategic initiatives (i.e., when and how customers are adopting technologies) and organizational initiatives (i.e., when and how employees and partners are adapting them) will be necessary.

Assimilation

At the other end of the spectrum, *assimilation* refers to the gap between how many organizations use technology and the laws and regulations that societies agree on to govern that use. Laws and regulation usually lag actual use, and this poses a different set of challenges for most companies. The gap between organizational use and regulatory frameworks is likely exacerbated for global companies that face differences in legal governance. For example, global companies have to deal with multiple legal and regulatory frameworks; policies that may work in one country may not work in another. Regulatory frameworks also vary across industries. One manager in a regulated industry actually points to regulation as a benefit for the company, as it provides clear guidance for what to do and not to do across all competitors in the industry. Some companies in regulated industries have reached out to regulators proactively to describe the innovations they hoped to enact, seeking guidance for how these initiatives should be implemented.

For most companies, waiting for the legal policymakers to catch up to practice is not an option. Organizations must adjust quickly enough to accommodate customer demand, while abiding by legal and regulatory guidelines. Indeed, among the major challenges facing rising stars such as Uber and Airbnb are the regulatory frameworks that a structure their behavior.

Adaptation

Between these two gaps of adoption and assimilation lies the most criti-cal gap facing nearly every organization today—adaptation. Adaptation is the gap between how the majority of individuals want (and expect) to use technology to engage with companies and how companies have adapted to support those interactions. Technological advances that people don't widely adopt pose potential strategic vulnerabilities in the future if a competitor figures out how to capitalize on them first. In the present, however, a disconnect between individual and organizational technol-ogy use represents a real competitive threat. If companies don't enable effective digital interactions with their customers, for example, then those customers can easily go to competitors or startups who will.

Fortunately, many companies recognize the need to engage with cus-tomers digitally, and this motivation is the driving force behind many initiatives. As companies embrace digital channels for reaching custom-ers, however, this effort can also exacerbate another facet of the adapta-tion gap—the space between employees and the companies for which they work. Employees are also customers of other companies, and they regularly experience streamlined digital interfaces for business inter-actions. Our data suggest that employees have become increasingly frustrated by the gap between what they are capable of accomplishing with technology in their personal lives and what they can get done at work when they are limited to email and nonmobile computing. As companies learn to engage with customers using technology, they often ignore their own employees. One executive at a company widely recognized as a digital leader notes that his employees could engage more easily with other companies as consumers using digital channels than they could with one another and with their own company. For example, it was often easier to apply for jobs outside the company than inside it. He noted that this technological disparity posed a potential talent problem.

The existential threat of this challenge for organizations cannot be overstated. Many of the most influential management thinkers have argued that firms exist solely because it is easier to do certain things within a company than outside it. Nobel Prize–winning economist Oliver E. Williamson, for example, asserts that firms exist to lower the transaction costs for certain types of exchanges that occur within the firm.[7] Strategy professor Robert M. Grant argues that companies exist to assimilate knowledge of employees.[8] A technological infrastructure that allows transactions and knowledge to flow more easily outside the company than within it threatens the very reasons that organizations exist. In many instances, the key problem companies face is the need to adapt quickly enough to address customer demands for digital interactions, while also changing the organization to meet the demands that technological advances stimulate in employees.

Improving Adaptation through Absorptive Capacity

The main problem posed by digital disruption is not the rapid pace of technological innovation but the uneven rates of assimilating these technologies into different levels of human organization. Thus, companies can effectively navigate the challenges of digital disruption by undertaking initiatives that are far more organizational and managerial than technical. Only by fundamentally changing the way the organization works—through flattening hierarchies, speeding up decision making, helping employees develop needed skills, and successfully understanding both opportunities and threats in the environment—can an organization truly adapt to a digital world.

In a foundational paper, published in 1990, Wesley M. Cohen and Daniel A. Levinthal introduced the concept of an organization's *absorptive capacity*. They define absorptive capacity as an organization's ability to identify, assimilate, transform, and use external knowledge, research, and practice. In other words, absorptive capacity is the measure of the rate at which a company can learn and use scientific, technological, or other knowledge that exists outside the firm. Cohen and Levinthal argue,

"The premise of the notion of absorptive capacity is that the organization needs prior related knowledge to assimilate and use new knowledge."[9] Expressed simply, this means that the more a company knows, the more it can learn. They note that "an organization's absorptive capacity will depend on the absorptive capacities of its individual members. . . . A firm's absorptive capacity is not, however, simply the sum of the absorptive capacities of its employees."[10] Absorptive capacity also depends on how an organization learns about the external environment and how different parts of a company transfer information to one another.

The obvious question is whether and how an organization can increase its absorptive capacity, with the goal of narrowing the adaptation gap. Cohen and Levinthal believe that absorptive capacity is an organizational competency that can be cultivated, and that firms can be purposeful about increasing their absorptive capacity. This question is also the focus of an article by Shaker Zahra and Gerard George published in the *Academy of Management Review* in 2002.[11] Collectively, the work of Zahra and George and Cohen and Levinthal suggests some specific steps that companies can take:

1. Expand talent diversity with a goal of increasing prior related knowledge. The challenge here, which we take up in chapter 9, is how to attract the right kind of individuals.

2. Augment the prior knowledge base of individual employees by providing them opportunities to develop skills for working in a digital environment.

3. Enhance the organization's mechanisms (e.g., sensing systems) for more effectively acquiring knowledge from the external environment, thereby increasing the firm's knowledge base.

4. Increase the velocity of internal information flows through initiatives that range from employee rotation to collaboration tools (e.g., Slack) to redesigned workplaces that encourage serendipitous exchanges among employees.

5. Focus on helping employees understand the "why" that Pfeffer and Sutton believe is so important to closing the knowing-doing gap.

Absorptive capacity builds on itself over time. Cohen and Levinthal argue that a company's existing related knowledge is a key antecedent for its ability to integrate new knowledge. This baseline knowledge is why companies maintain research and development initiatives, rather than just purchasing the innovations themselves. In other words, the organization needs to learn how to learn. Conversely, if the organization kills innovative experimentation for a time, integrating new knowledge in the future can be more difficult.

Adobe: Combining Employee and Customer Experience

The software company Adobe is attempting to address the technological disparity between customer and employee by uniting their experiences under a single leader. In 2012, Adobe undertook a major overhaul of its business: It shifted from software sold through long-term licenses and shipped in shrink-wrapped boxes to a subscription model in the cloud, with lower monthly fees designed to attract a larger portion of the market. The company also added digital marketing to its portfolio, expanding from its traditional focus on the graphic design and publishing industries.

Adobe's senior management realized the importance of culture, talent development, and employee engagement in making such fundamental changes to its structure and business model. Although the company gained new digital talent through a series of acquisitions, Adobe's leaders strongly believed that culture and talent development efforts had to be strongly fused. To foster that connection, the company took the bold step of putting employee and customer experience under the same organizational umbrella and leader. "We have had a long-standing commitment to investing in our employee experience," says Donna Morris, executive vice president of customer and employee experience. "But we felt we needed to change our culture and put the same emphasis on customers. We wanted all employees to share the common perspective that they do contribute to the customer experience."

As an example of the combination, Morris points to Adobe's Experience-athon program. Designed to put employees in their customers' shoes and spur change, the program turns employees into product users who provide immediate feedback to the customer organization, Adobe. "There is such an opportunity to have our employees experience our products and services firsthand before we offer them to customers," she says. "By combining employee and customer experiences, we are able to create rich customer experiences through high levels of employee engagement."

Takeaways for Chapter 2

What We Know	What You Can Do about It
• The key challenge in responding to digital disruption is the ever-increasing gaps between the rates at which the individual, company, and society adapt to technological change. • **Absorptive capacity** is the rate at which organizations can identify and effectively assimilate knowledge and innovations. It is a learned capability that can build on itself over time.	• Invest in five (or more) digital "field trips," in which you take a close look at what digital leaders actually do in terms of deploying technology for internal processes and customer interactions. Share your observations with colleagues and invite them to add their observations about what digital leaders are doing. • Based on observations from the field trips, identify at least three digital pilots you can launch in the next month that will give your organization an opportunity to develop new capabilities. Make sure that your pilot plan includes clear learning objectives that you can assess at the end of the pilots. • Execute, evaluate, and repeat at regular intervals.

3 Moving beyond the Digital Transformation Hype

In the past two decades, the term "enterprise transformation" has become widely used by organizations seeking a way to describe a radically new way of operating. Variants have followed, ranging from finance transformation to supply chain transformation to marketing transformation. As technology has increasingly disrupted the status quo, we have seen a shift from the term "enterprise transformation" to the more au courant one, "digital transformation." "Transformation" is a powerful term, conveying a sense of dramatic (as opposed to incremental) change. Yet, after surveying and interviewing thousands of organizations and individuals over the past six years of research, we have come to believe that the term "digital maturity" may be much more helpful for organizations seeking to understand how to engage effectively with a fast-moving and continuously changing environment.

Digital Maturity: Becoming a Digital Organization

In this book, we forward an aspirational goal of digital maturity to suggest that a balanced approach to digital disruption is needed to deal effectively with the changes at hand. We still use terminology like "digital disruption," "transformation," and similar phrases throughout this book simply because most people are currently using this language to describe the trends that are the topic of our research. Nevertheless, we also want to move past these terms to describe the type of meaningful, comprehensive, and lasting changes necessary and possible for organizations, talent, and leadership to successfully adapt to a competitive environment increasingly defined by technology. We hope to avoid much of the hype often associated with digital trends.

We define digital maturity as

aligning an organization's people, culture, structure, and tasks to compete effectively by taking advantage of opportunities enabled by technological infrastructure, both inside and outside the organization.

This definition draws on established organizational theory developed by David A. Nadler and Michael L. Tushman (1980).[1] They spearheaded the idea of organizational congruence as a primary ingredient for optimal corporate performance. The concept holds that only when the essential components of a business—its culture, people, structure, and tasks—are tightly aligned can the company achieve powerful results. For example, a conservative and hierarchical organization that recruits energetic entrepreneurs may not be able to harness their drive and energy. Similarly, an organization that has completely flattened its structure may struggle if its culture shuns risk.

At first blush, the elements of congruence may seem intuitive or even old hat. But in a constantly changing landscape, the concept of congruence gains new meaning and currency as many companies fail to ensure that all these organizational elements are firing together over time. Although adopting digital platforms is necessary for the alignment needed in the emerging competitive digital environment, the use of such platforms alone is not sufficient for digital maturity.

"I'm often surprised by how detached executive management is from the actual culture of their companies," says Chip Joyce, cofounder and CEO of Allied Talent, a Silicon Valley–based boutique management development consultancy. "C-level executives often portray their organizations as transparent, open to risk taking, and having high morale. But as you move down the organizational structure, managers rarely believe it and say that the level of trust is very low." To navigate the complexity of contemporary business, companies should align their culture, people, structure, and tasks with one another and with the digital environment so that executives can effectively address the

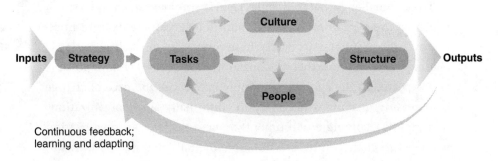

Figure 3.1
Source: D. A. Nadler and M. L. Tushman, "A Model for Diagnosing Organizational Behavior," Organizational Dynamics 9, no. 2 (1980): 35–51.

challenges of a constantly changing landscape (figure 3.1). Throughout this book, we explore how executives are creating digital congruence and driving substantive change across their organizations.

Nadler and Tushman's theory, however, did not account for the evolution of technology. In figure 3.1, we've added a circle around the four essential elements of a firm to account for the changing digital landscape. An organization's culture, people, tasks, and structures all occur within a given digital environment, which interacts with these elements; they need to be realigned as this environment changes. This change happens through experimentation and iteration, a topic we deal with in depth in chapter 14, as the organization recalibrates its components to match the capabilities in and characteristics of the current digital environment. This process of experimentation and iteration is represented in the feedback loop from the organizational outputs to the business strategy. As new technologies are introduced, adopted, and encouraged to mature, organizations sense the changes wrought by these technologies and adapt themselves accordingly.

Put simply, we believe that the fundamentals of good management haven't become completely irrelevant just because of digital disruption. Digital disruption has simply changed the conditions under which those management principles operate, and they will play out differently as a result.

The key challenge that companies face is twofold. First, many have not realigned their organizations to account for developments in technology over the past ten years or so. These companies must close considerable gaps between how their organizations operate and what is possible and expected in the current environment. Second, the pace of change is increasing, making it necessary to continually adapt organizations to a rapidly evolving environment. Some aspects of an organization will need to change as it becomes more digitally mature and learns to adapt more quickly, but others won't. You don't need to throw out the playbook of good management to mature digitally, but you do need to update that playbook for the current environment. Understanding how to do so—what needs to change and what doesn't—is one of the primary objectives of managers in a digital environment.

Thus, our concept of digital maturity should not be misunderstood as a staid end state of the organization. Rather, digital maturity—or, perhaps more accurately, digitally maturing—is a flexible process by which the organization can continually adapt to a changing technological environment, realigning its people, culture, tasks, and structure in response. This challenge may seem daunting, if not impossible, for traditional organizations. In fact, without significant changes, it may be. Leaders may need to rethink how their organizations function and then develop new talent models, cultural characteristics, task definitions, and an organizational structure that are more amenable to this fluid environment.

What Is Different, What Is Not, and How to Tell the Difference

Our view that any organization can rise to the challenges of digital disruption is in contrast with the prescriptions of many so-called gurus who want you to believe that only a select few can truly think digitally. They want you to believe that the path to successful digital transformation requires secret knowledge, either naturally instilled in the millennial generation from birth (i.e., the digital natives) or possessed only by a select few who have worked at Silicon Valley technology firms. They want you to believe that companies without large dosages of this secret knowledge will be locked out of an abundant future.

The ancient Greeks developed a philosophical concept called *gnosis*, which referred to a secret knowledge that allowed its possessors to achieve enlightenment. This secret knowledge was possessed only by a select few, who occupied a privileged position above those who did not have it. Being an effective digital leader *does not* require secret knowledge. It *does* mean leading your organization amid changing environmental conditions that you may have little control over. You may find yourself thinking, "Isn't this what being an effective organization has always been?" We get this response often. In short, the answer is "yes." Many of our prescriptions for achieving digital maturity won't be new, but we are shocked at how often these basics seem novel to some managers and how often managers forget the basics in the face of digital disruption.

The essentials are easy to lose sight of in an area that changes so quickly and is perpetually focused on the latest and greatest. The ride-sharing platform Uber was launched as recently as 2009 (as UberCab), and the enterprise social media platform Slack started the same year, yet both have significantly influenced how work is done. As recently as 2012, companies were asking whether Facebook could successfully transition to mobile environments. Today, nearly 80 percent of use and 70 percent of Facebook revenues come from mobile platforms, and many companies now assume the mantra of "mobile first."

Walmart: Digital Maturity and the Long Game

In 2016, Walmart made headlines by acquiring online retailer Jet.com for more than $3 billion, then bolstering that purchase with other e-commerce acquisitions, including Zappos competitor Shoebuy.com, women's fashion brand Modcloth, and outdoor outfitter Moosejaw. Just a few years earlier, Walmart had begun playing the long game, looking at a ten-year horizon of investments to strengthen its own digital capabilities. Its leaders realized that competing in a digital world demanded more than snapping up leading

(continued)

online retailers to compete with Amazon.com. Walmart is rethinking virtually every aspect of its business to keep pace with changing customer behaviors over the long haul.

First, Walmart is revamping its business strategy, but not just in terms of the next steps it can take today, which the company is doing with everything from developing shopping apps and making investments in frontline training to strengthening its logistics muscle. Beyond these steps, the global retailer is creating a digital strategy to address what its leaders believe the company must be able to do ten years from now. As Jacqui Canney, Walmart's executive vice president of global people, puts it, "Between the five- and ten-year mark, people are going to shop in very different ways and expect different experiences. That's why we're focused on positioning ourselves for the future."

Achieving its vision of a digital future requires the company to work in fundamentally different ways. Enterprises need talent, organizational structure, and culture to be in sync with digital environments around them. "Not everybody sees digital as changing the way you work," says Canney. "It's about educating your people that it's not just about technology."

Walmart is pursuing a score of nontechnical, operational changes across the organization to prepare for this digital future. The company is bringing digital talent skills to the entire organization, including a customer-first mindset, collaboration, and design thinking. These changes are also an integral part of executive job descriptions. "Metrics and objectives focused on digitizing the businesses are now required," says Canney. "This year we've added digital leadership as one of the core competencies our leaders must have in order to be promoted."

Why Digital Maturity?

Shifting focus from digital transformation to digital maturity has several benefits for organizations seeking to adapt to an increasingly digital competitive environment. Our choices in terminology and definition are drawn from the psychological definition of maturity. In a paper published in the *International Journal of Humanities and Social Science Invention*, Rita Rani Talukdar and Joysree Das define maturity as "the ability to respond to the environment in an appropriate manner. This response is generally learned rather than instinctive."[2] Maturity has five elements relevant to digital environments:

1. **Maturity is a gradual and continuous process that unfolds over time.** Just as no person becomes more mature overnight, no organization can become digitally mature overnight either, regardless of what vendors may suggest. Individuals encounter different developmental goals at different stages of their lives—from toddlerhood into adolescence, adulthood, and retirement. Similarly, companies may experience different challenges at different stages of their development, and they can always continue to grow and adapt to become even more digitally mature.

2. **Gradual maturation should not be confused with less significant changes.** The difference between toddlers, children, teenagers, and young adults is substantial, even though it takes time for these changes to be fully realized, and the changes may not be obvious from day to day. As your company becomes more digitally mature, you may find that you need to do business in very different ways. Just as toddlers, teenagers, and adults engage in the world in different ways, so might digitally maturing companies change accordingly.

3. **Organizations may not fully know what they will eventually look like when they begin to mature.** Only a small percentage of children who want to grow up to be firefighters, cowboys, or princesses actually do so, yet that lack of perfect knowledge about the destination doesn't keep the maturation process from happening. Even though many organizations cannot describe what a digitally mature version of themselves will ultimately look like, it shouldn't stop the process from beginning. *Solvitur ambulando*, a Latin phrase attributed to the Greek philosopher Diogenes of Sinope, means "It is solved in the walking." You may have a better idea of what digital maturity is for your company only once you start moving toward it. In fact, learning more about the environment and testing one's place in it are important parts of maturing.

4. **Maturation is a natural process, but it will not happen automatically.** Digital maturity is the process of your company learning how to respond to the emerging competitive environment appropriately. Yet, your organization, leaders, and employees will not instinctively know how to do this. Not even the digitally native millennials necessarily know how to apply their skills in an organizational context. Individual use of technology does not always translate into knowing how to use these tools effectively within or as an organization. Managers must develop working knowledge of digital trends to lead their organization to adapt in the right ways. Conversely, not responding appropriately to the environment is unnatural. We have all likely known immature people. Certainly, most managers don't want their company to be that type of organization.

5. **Maturity is never complete.** While it may be tempting to assume that digital maturity is limited to young Silicon Valley companies, even established companies and employees can exhibit the changes necessary to adapt. In fact, one of the most refreshing trends we have seen over recent years is the increase in large legacy companies making the types of changes necessary to adapt to a digital world. They may not all be successful or make the right types of changes to survive, but their willingness and ability to change bodes well for the future of at least some of these companies. The question is whether other companies are willing to attempt these changes rather than letting denial and fear of failure prevent them from taking action. It is never too late to begin becoming more digitally mature, and the process is never complete.

Moving beyond the Hype

Terms such as "digital disruption" and "digital transformation" have been tossed around so often that they have lost most of their power. Management gurus and self-declared futurists have been proclaiming the danger so loudly and for so long that many people have ceased paying attention, much like the response of the townspeople in Aesop's

tale "The Boy Who Cried Wolf." Several executives we
actually avoid using digital terminology precisely because
in the marketplace. "It sounds like the catchy new phrase, a
people who react poorly to that," says Christine Halberstadt, vice president, asset management at Freddie Mac. "People should speak in words
that resonate. There is strong support on the part of our management
team for digital transformation, but I think they are just now getting
used to that phrase." An executive at another organization quips that
his company goes so far as to "trick" its employees into embracing digital transformation by couching it in nondigital language.

Although we are wary of abandoning the term "digital" entirely
(mainly because we fear that no one would read our research reports
and book), some clarification is likely necessary before we continue.
Merriam-Webster defines "digital" as "characterized by electronic and
especially computerized technology." Most important, it defines the
word as an adjective, yet it is increasingly used as a noun. People
often say, "I'm in charge of digital." "We're going digital." This distinction between "digital" as a noun and as an adjective is important. A
noun has a distinct identity of its own, but an adjective modifies a noun.
So "digital" doesn't have a distinct identity of its own, but it changes
other nouns. Companies can have digital marketing, digital strategy,
digital talent, digital leadership, digital culture, and more. The presence of the modifier "digital" focuses on how these business processes
need to change as they are increasingly characterized by technology.
We encourage readers, however, not to get hung up on the particular
choices of terminology used here or elsewhere, but to think about the
larger issues at play. Digital maturity by any other name would be just
as effective—and maybe more so, depending on your environment.

Measuring Digital Maturity

While our definition of digital maturity is an important conceptual foundation to our work, how we actually measure an organization's digital
maturity is also important to describe. In each of the four years of our

survey, we asked respondents to "imagine an ideal organization transformed by digital technologies and capabilities that improve processes,
engage talent across the organization, and drive new value-generating
business models now and in the future." We then asked respondents to
rate their company against that ideal on a scale of one to ten, with one
being as far as possible from that ideal and ten being a perfect match with
the ideal.

We group companies into three categories: early (rated one to three
by respondents), developing (rated four to six), and maturing (rated
seven to ten). The dispersal of companies along this spectrum resembles
a normal distribution, with just under half of respondents placing their
organizations in the middle, developing category; a quarter of respondents categorizing their companies as early; and just over a quarter rating
their firms as maturing (figure 3.2). Interestingly, while these numbers
have remained stable for a few years, we have seen a shift toward digital maturity in our most recent study. Does this mean that companies

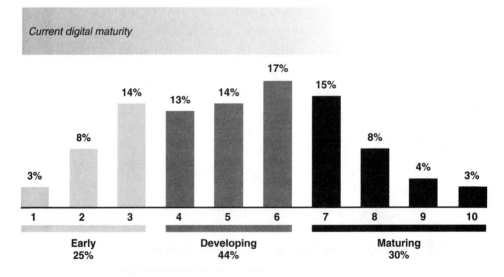

Maturity percentages do not total 100 due to rounding.

Figure 3.2

are beginning to make progress toward digital maturity? Early evidence, both quantitative and anecdotal, suggests that this may be the case.

This way of measuring digital maturity has several strengths and weaknesses. Most important, it relies on perceptual data. Although we previously covered some of the limitations of perceptual data, these limitations bear repeating, given the importance of these data to our research. Do the average employees know what the "ideal" digital organization looks like, and are they adequate judges of how effectively their company has achieved that ideal? Maybe, and maybe not. Much like our earlier discussion of whether respondents know what constitutes an "adequate" response to trends, regular employees may actually provide better insight than top-level managers, because they know how digital efforts are changing the way work is done (or not) at the granular level.

Furthermore, by asking respondents to compare their organizations to a perceived ideal, we were not asking them to assess how digitally mature their organization was but—rather—how digitally *im*mature it was. Much as Supreme Court Justice Potter Stewart couldn't define obscenity (i.e., "I know it when I see it"), average employees can't necessarily define digital maturity, so we asked respondents simply to identify it if and when they saw it.[3] By and large, our respondents aren't seeing it in their own companies. They recognize that more is possible.

Another advantage of our approach to measuring an organization's digital maturity is that it accounts for the likelihood that not all parts of an organization are equally digitally mature or maturing at the same rate. Indeed, when we presented some of our results at an executive education seminar, we asked participants to evaluate the maturity of their own companies. Three different respondents who worked at the same large company reported that it was in each the early, developing, and maturing groupings. When we pushed them further on this response, it became clear that digital maturity was not uniform across the organization. The corporate real estate division was using advanced analytics to determine and optimize building utilization rates, the recruiting function was just beginning to experiment with new ways to manage the

talent pipeline, and other parts of the organization were barely thinking about the effects of technology on business. The approach of asking
individuals to rate their own experiences of their organization's digital
maturity allows us to explore the particular organizational characteristics associated with digital maturity, without having to debate the particular merits of whether a specific organization is digitally mature or
not. So, although we will speak of "digitally maturing" organizations
throughout the book, what we really mean are the characteristics or
parts of an organization with which employees engage in their work.

In an effort to isolate the key cultural approaches to digital maturity, we performed a statistical approach called "cluster analysis" on
certain organizational attributes completely unrelated to our maturity
measure. What emerged was a three-cluster solution that nearly per-
fectly matched the maturity groupings described here. (This analysis
is discussed further in chapter 11.) So, independent data and analysis
provide evidence that these three groups are, in fact, the best way to
interpret our data. This tripartite grouping has also been received well
by interview subjects and groups to which we have presented our findings. So, while we confess that our measure and groupings may not be
perfect assessments of digital maturity, we have found both to be supported by a separate analysis of the data and helpful ways of talking
about the technology issues facing most organizations. They provide
a way of describing key differences in organizations' digital maturity
and help organizations strategize about how to become more digitally
mature.

The Importance of Humility

One final distinction in our grouping warrants further discussion. Consistent with our understanding that digital maturity is never complete,
we intentionally refer to the last category as "maturing" rather than
"mature," reflective of that changing nature of the environment. The
rate of change that organizations respond to will not abate within the

foreseeable future. Technology will keep changing faster than individuals, who will keep adapting to that change faster than organizations. Indeed, you will likely notice an interesting trend in our data throughout the book, one that we observed in all four years of research. That is, many of the optimal digital maturity results tend to peak at around level eight, gradually tailing off somewhat for respondents that rate their companies at levels nine and ten compared to the ideal. Many reasons could explain that dip at the higher levels of maturity, including small sample sizes in those respondent groups.

On the other hand, a healthy recognition that your organization falls—and will always fall—a little short of the ideal is an important element of maturity. Respect for the nature of ongoing technical change and recognition that there is still room to get better may in fact be the pinnacle of maturity. Even digitally maturing companies still struggle with maintaining their maturity. If leaders have to make conscious efforts to continue maturing digitally, then the rest of us might be well served by adopting a similar approach. Thus, companies will need to develop processes that allow them to adapt to those constant changes. We describe this process of current change and adaptation as "maturing," rather than "maturity" to reflect this reality. Reid Hoffman echoes this sentiment in his book *The Start-Up of You,* admonishing individuals to be in a state of "Permanent Beta"— meaning that they should be constantly experimenting and adapting to accommodate these changes in the environment.[4] Digital maturity is about continually realigning your organization and updating your strategic plan to account for changes in the technological landscape that affect your business.

More Opportunity than Threat

Another way that companies exhibit a lack of humility is that they tend to view digital disruption in purely optimistic terms. We asked respondents the extent to which their companies viewed technology as

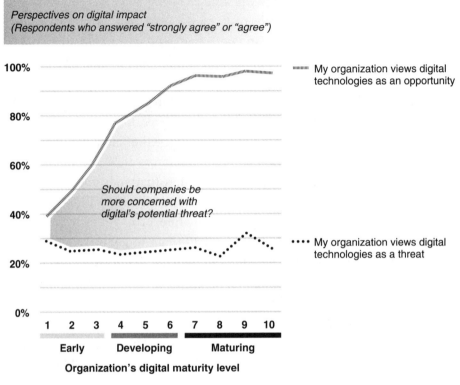

Perspectives on digital impact
(Respondents who answered "strongly agree" or "agree")

- ▬▬▬ My organization views digital technologies as an opportunity
- •••• My organization views digital technologies as a threat

Should companies be more concerned with digital's potential threat?

Early   Developing   Maturing

Organization's digital maturity level

Figure 3.3

an opportunity or a threat to their organization. The disparity between the responses is striking (figure 3.3). While more than 80 percent of respondents reported that their organizations viewed technology as an opportunity, only 26 percent viewed it as a threat. While viewing technology as an opportunity rises with digital maturity, viewing it as a threat is stable across maturity levels. This striking disparity is logically inconsistent and represents a naïve optimism on behalf of business leaders, reminiscent of Voltaire's character Professor Pangloss, who believed that he lived in the best of all possible worlds. But if technology represents an opportunity for your organization, it also represents a threat for your competitors—and vice versa. This optimistic view as a reason for not responding more aggressively is a bad one.

Similar naïveté is reflected elsewhere in our data. We asked respondents whether they expected demand for their organization's core products or services to increase or decrease because of digital trends in the next three years. Over two-thirds of respondents said that they expected demand to increase, while only one-tenth reported that they expected demand to decrease. It's certainly possible that the digital tide will raise all boats by expanding markets and increasing buying power. Indeed, these companies may all be headquartered in the mythical town of Lake Wobegon, "where all the women are strong, all the men are good looking, and all the children are above average." The likely reality, however, is that these respondents are simply underestimating the odds that the same digital trends will also lift the fortunes of their competitors by the same or greater amounts. A digitally mature mindset recognizes that digital disruption represents both an opportunity for and a threat to your organization and responds appropriately.

Transformation on the Fly at The Atlantic Monthly Group

Market conditions often foil even the best-planned strategies or iterative experiments to move an organization forward. That's when a traditional business is forced to turn its disadvantages into a competitive advantage. Kimberly Lau, EVP of digital and head of business development for the *Atlantic*, recalls the magazine ten years ago, before she worked there, as a struggling legacy media company that had to "pivot to digital" when its core print business was in jeopardy.

"The running theory at the time was that print would be dead within five years," so seeking immediate digital options became imperative to the venerable publication. "There was no scenario where leaders could just walk away from that revenue." The *Atlantic's* experimentation with bloggers and online content as early as 2008 helped strengthen the brand and its content today, Lau says.

The need for speed created urgency and made the company less risk averse; it had to become agile on the fly, she says. In fact, the print magazine is a thriving—though smaller—part of the business today, and the media industry has moved to more paid-subscription and multimedia content options.

Takeaways for Chapter 3

What We Know	What You Can Do about It
• **Digital maturity** describes how your organization adapts to and narrows the gaps in engaging appropriately with a fast-moving and rapidly changing digital environment.	• Take the time (preferably with colleagues) to visualize what a more digitally mature version of your organization might look like. Fuel with insights from this book, your digital "field trips," and additional research.
• Digital maturity involves aligning an organization's people, culture, structure, and tasks so that the organization takes advantage of the opportunities enabled by technological advances both inside and outside the firm.	• Look across the organization broadly to identify which business units, departments, teams, or functions come closest to delivering the vision and which are the furthest away.
• Digital maturity is a moving target, continually changing along with technology. Therefore, we refer to the most digitally advanced companies as "maturing," recognizing that the process is continually evolving.	• Devise an action plan that would help one of the laggards leapfrog its current state. Your action plan should address what process, talent, technology, and operating principles need to change, as well as which ones need to remain constant.
	• Keep on tackling each organizational laggard until the gap between worst and best is narrowed.

4 Digital Strategy for an Uncertain Future

Our four years of research consistently highlight that a clear and coherent digital strategy is the single most important determinant of a company's digital maturity. More than 80 percent of respondents who rate their companies as digitally maturing asserted that their organizations have a clear and coherent strategy compared with 15 percent of respondents from the least mature companies (figure 4.1). This finding raises the question—what exactly is a digital strategy?

Entire books have been written on digital strategies, and many of these are quite good. George Westerman, of MIT; Andrew McAfee, also of MIT; and Didier Bonnet, of Capgemini Consulting, offer a nice treatment in their book *Leading Digital* of how large companies use technology to gain strategic advantage. They use the term "digital master" to describe this transformation.[1] McAfee partnered with another MIT colleague, Eric Brynjolfsson, to produce the best seller *Machine, Platform, Crowd*, about how artificial intelligence, social media, and blockchain are creating new opportunities for doing business.[2] David Rogers, of Columbia University, penned a volume, *The Digital Transformation Playbook*, on how leaders should update their thinking. He argues that digital forces are disrupting five key domains of strategy: customers, competition, data, innovation, and value. Reminiscent of Michael Porter's *Five Forces*, he argues that digital disruptions require new ways of thinking that ultimately allow companies to adapt their value proposition for the digital age.[3] Marshall Van Alstyne, of Boston University and MIT, and his colleagues wrote a book, *Platform Revolution*, on strategies just for platform

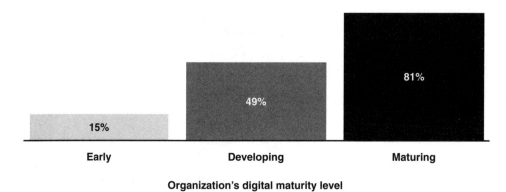

Chapter 4

*Our organization has a clear and coherent digital strategy
(Respondents who answered "strongly agree" or "agree")*

81%

49%

15%

Early   Developing   Maturing

Organization's digital maturity level

Figure 4.1

businesses. They contrast platform businesses, such as Facebook, Apple Inc., Amazon, and Microsoft, with more traditional pipeline businesses that rely on typical supply chain concepts of moving products to market. They outline how the economic rules of platform businesses differ from those of pipeline businesses.[4]

We can even recommend John Gallaugher's undergraduate textbook, *Information Systems: A Manager's Guide to Harnessing Technology*, as perhaps the most consistently up-to-date discussion of digital strategy out there—he updates the material yearly, and it's readily available online as a PDF file. Gallaugher (a Boston College colleague of coauthor Kane) focuses on a more traditional view of digital business, discussing how and why specific companies like Google, Amazon, and the like are reshaping the business landscape.[5]

The aforementioned books are insightful and well written, by authors and author teams with strong credentials and expertise. They are likely to be helpful for the manager seeking to understand digital strategy better. These books share a common characteristic—they are mostly retrospective. They examine successful cases of digital strategy, analyze why they were successful, and present the success principles that managers should follow. These retrospective lessons are invaluable when your company faces business problems precisely like those faced by Amazon, Facebook, Zara, or Google. Managers can certainly benefit from the "lessons learned" when their companies share the strategic challenges and opportunities represented by the case studies. Incidentally, there are also plenty of other books on digital strategy (Amazon seems to list thousands), many of which we don't find particularly substantive or helpful.

Many of the strategic challenges and opportunities that your company faces from digital disruption may be unique, with nuances related to your company, your industry sector, your geography, or your competitive environment. The nuances may overwhelm the similarities, and consequently, opportunities to "lift and shift" digital strategies may be quite limited. This issue is not limited to digital strategies. Management literature is full of articles and books that study successful companies and exhort others to emulate them. Unfortunately, some of these successful companies go on to flounder.[6] We suspect a key reason many of these strategies were successful to begin with is that the leaders saw opportunities or challenges before others did. Thus, although we agree that leaders need to examine past successes and failures of others to make informed decisions, we think that the greater challenge in forming a digital strategy is actually thinking proactively and identifying the strategic moves that are unique to your situation. You don't want to be fighting the last war but preparing for the next one.

Since the primary challenge that organizations face is adapting themselves to a changing environment, a company's digital strategy will necessarily evolve as the environment does. Therefore, digital strategy is not necessarily a singular long-term plan to which the organization doggedly adheres and that it executes over a multiyear timeframe. Rather, it is a recursive process of identifying the overall goals of digital business, developing short-term initiatives that get the organization closer to the goal, and then rethinking the nature of those goals based on what the organization has learned from those short-term initiatives.

Lack of Strategy as a Key Barrier to Digital Maturity

We asked respondents what was keeping their companies from maturing digitally (figure 4.2). The barriers that respondents reported differed by maturity levels. Perhaps the most significant barrier that plagues most companies regardless of maturity level is too many priorities. This was the top barrier for both developing and maturing companies, and the second most important barrier reported for early stage companies. Yet, this is also essentially a strategy barrier. When organizations have too many priorities, they do not have a good idea of what their strategic focus actually is.

The legendary Stanford Business School professor James G. March described this tradeoff between established and new business initiatives as exploration versus exploitation.[7] Exploration processes are about innovation, often resulting in lower-performing short-term outcomes required to figure out the new processes but better long-term outcomes that result from the new processes. Exploitation, in contrast, yields better short-term outcomes from organizations doing things in familiar ways but lower overall long-term outcomes by not searching for new ways of doing things. Most of the exploration-exploitation literature advocates for a balance between established and innovative processes to maximize both long- and short-term performance. Charles A. O'Reilly and Michael L. Tushman refer to organizations that effectively balance the needs for exploration and exploitation as "ambidextrous organizations."[8]

We deal with the other key barriers throughout the book. For example, the lack of a strategy is an important barrier for both early and

	Early	Developing	Maturing
Top barriers by maturity stage	1. Lack of strategy	1. Too many priorities	1. Too many priorities
	2. Too many priorities	2. Lack of strategy	2. Security concerns
	3. Lack of management understanding	3. Insufficient tech skills	3. Insufficient tech skills

Figure 4.2

developing companies, and developing a digital strategy is the focus of the next chapter. Likewise, insufficient tech skills are problems for both developing and maturing companies, which is a topic we deal with in chapter 7. We hope that this book addresses the "lack of management understanding" from which early stage companies suffer. The only barrier that we do not deal with in this book is security concerns, which are elevated to a higher priority in maturing companies. We highlight this barrier simply as evidence that maturing digitally likely introduces new problems that companies have to deal with at later stages of development.

MetLife's Digital Lens on Business Strategy

A global financial services company facing intense competitive pressure from its global peers and fintech upstarts, MetLife develops strategies that address the financial services it provides and how technology can make them more competitive.

"There are four pillars that make up MetLife's strategy which aren't of and in themselves digital," says Martin Lippert, executive vice president and head of global technology and operations. "The first is optimizing value and risk. The second is delivering the right solutions for the right customers. The third is strengthening our distribution advantage. And the fourth is driving operational excellence."

Lippert wants the organization to think of these pillars with a digital mindset to improve the customer experience and make the organization as efficient and effective as possible. As he puts it, "Digital sits in the middle of these four pillars and brings them all together."

To imbue the company with a digital mindset, Lippert works both top down and bottom up. To strengthen digital thinking in senior management, he recently took several top executives and other senior leaders to Silicon Valley to meet with venture capital executives in whose firms MetLife has invested. The company has also partnered with technology companies, start-ups, and universities to bring new ideas and approaches into executive offices.

To drive ideas from the bottom up, MetLife hosts an annual event called MetLife Ignition. At the company-wide gathering, portfolio companies from the venture capital firms MetLife has invested in present ideas to employees, including describing the challenges that the innovations address. The event stimulates many new ideas that often move into proof of concept and, if preliminary results hold, are eventually rolled out globally.

Think Differently to Develop Digital Strategy

Developing a digital strategy does not necessarily involve a bunch of smart, tech-savvy people sitting in a room thinking up a grand design that then gets implemented over a multiyear timeframe. Since successful initiatives often yield new possibilities for doing business (see chapter 5 for more), developing digital strategy is a recursive process involving three steps in a repeating loop:

- **See differently.** This step involves managers making sense of the actions possible in the current environment. They scan the environment for technological and organizational capabilities and determine a single action that will yield the biggest positive impact on the organization. The action with the biggest impact may be to eliminate some of the barriers to effective digital strategy. The single action that managers identify will be the strategic goal driving the next steps of the process.

- **Think differently.** The short initiative in the previous stage may or may not be successful in surfacing a single strategic goal. If it is successful, leaders should consider whether new capabilities may be possible as a result of working toward this goal. If they could not identify a strategic goal, leaders should take time to figure out why not and how those reasons might affect other efforts to develop digital strategy, repeating this step until they can identify a goal.

- **Do differently.** In this step, the organization plans a six- to eight-week initiative to make meaningful progress toward their strategic goal. Significant resources are leveraged for the organization to attempt to work differently in this short period.

- **Repeat.** The leaders take the knowledge developed in the final stage and reassess organizational opportunities in light of this new knowledge, repeating the cycle.

This process of digital strategizing is reminiscent of orienteering, the skill of getting from one place to another using a map and compass. In this process, you are constantly taking your current position based on features of the environment and charting the best path to your destination based on that location. The key difference in digital strategy, however, is that the goal to which you are traveling keeps shifting. Thus, effective digital strategy is really an ongoing process of strategizing that keeps the organization moving toward this evolving goal.

Think Long Term . . . and Then Think Longer

The need to continually revisit your digital strategy does not limit your organization to short-term thinking. Quite the opposite. Unless you have a good sense of where you are heading, your short-term objectives could lead you in the wrong direction. Deloitte's John Hagel laments that most companies don't look far enough ahead when thinking about digital strategy. Instead of the one- to three-year time frame that most companies use for digital strategy, Hagel advocates using a ten- to twenty-year timeframe in addition to these short-term goals.[9] In our own research, only 2 percent of respondents said their company thinks about digital strategy on such a broad time horizon, and only 10 percent think five years or more in advance.

It may seem crazy to do strategic planning on such a long horizon. Few people can adequately predict what digital trends will dominate in the coming years, let alone the coming decades. After all, who could have predicted the current state of mobile, social, and analytics tools in the mid-1990s, before the dot-com boom even got started? Yet the trends started then have played out according to a generally predictable path, even if some of the specific outcomes—such as which companies would prevail as a result of these trends—were unforeseen. While a smaller number of companies than expected profited directly from those trends, many more have been disrupted by them.

More and more leaders are recognizing that such a long view is necessary. For example, we were surprised to learn how long term Walmart's digital strategy is, as described in the previous chapter. Walmart recognizes that they need to start their digital transformation now, because what customers will want ten years from now will be radically different from what they want today. Maturing companies are roughly twice as likely as early and developing companies to look five or more years out when developing their digital strategy. This long-term vision helps leaders better see the current environment in relation to its future state. It helps managers figure out the most productive ways that they need to begin to act differently to ready themselves for this future environment. Yet, how does a company engage in this process?

The management scholar Benn Konsynski, of Emory University, recommends that organizations begin by "reverse engineering the future." Instead of considering the next strategic steps made possible by today's technology, he recommends imagining what the future technological infrastructure will look like and only then plot the next steps to get you to that future. Otherwise, your short-term initiatives may just take you a step in the wrong direction.

Cisco: Getting Ahead of Digital Disruption

The networking company Cisco is getting out ahead of digital disruption. Although it's tempting to think of technology companies as disruptors themselves, they can also be disrupted as technological trends shift. One needs to look no further than former giants such as Digital Equipment Corporation, Wang Computers, and Nortel Networks to know that the technology industry is also susceptible to disruption. In fact, Facebook has taken over the old Sun Microsystems campus for its Menlo Park Headquarters and still keeps a Sun Microsystems sign up to remind employees how quickly things can change.

Although Cisco was still a highly profitable company, its leaders recognized that the world was changing. James Macaulay, senior director in the Cisco Digitization Office, notes, "Cisco is not a company that leaps to mind that says we need to transform. But we know it's coming. We know it's coming in terms of our competitive landscape and we intend to be as successful in the next ten years, indeed more successful, than we were in the preceding ten. Our customers very clearly are saying they want more flexibility in terms of how they consume technology. They want different types of business outcomes. They are, themselves, facing all these new competitive pressures as a result of digital disruption. They feel a tremendous sense of urgency. We know that we need to change our business model and how we operate to meet our customers' expectation in the years ahead.

"Historically, Cisco is probably best known as being a hardware company. The world very clearly still needs a lot of hardware, but we also know that because customers want to consume differently, and in some cases, they want to consume as a service. One of the biggest changes that we're driving at Cisco is really around our business model; moving more toward a recurring revenue-type of model, but also in terms of our portfolio, expanding our core business around hardware to include a broader revenue mix of software and services.

"It's really about changing our customer experience, how they buy from us, how they consume our technology, changing the business model, the route to market and all the operations that go along with delivering, selling, supporting, maintaining our customers' technology. It begins with the customer, then the business model and then, the operational changes." More companies need to think like Cisco, about how the competitive environment will change in the coming decade, and begin to rethink their businesses now to be ready for that change.

A Thought Exercise for Long-Term Strategic Thinking: Autonomous Vehicles

Conducting a thought exercise can help highlight the benefits of long-term strategic planning about digital trends.[10] A good example of how and why to adopt this long-term time frame is the market for autonomous (a.k.a. self-driving) vehicles, which is likely to become mainstream over the next decade or so and has strategic implications for a wide range of industries beyond automotive, ranging from property and casualty insurance to health care and real estate. It may be difficult to predict accurately *exactly* when or how autonomous vehicles will become mainstream. It's safe to say, however, that this future will eventually become reality in the next ten to twenty years. Ubiquitous autonomous vehicles may affect a diverse set of industries:

1. **Auto dealers.** Automobile dealers would be significantly affected by the widespread adoption of autonomous vehicles. When cars can drive themselves, they don't need to wait for their owners to pilot them. Passengers could subscribe to an Uber-like service that would be dispatched to pick up passengers on demand. Individual people

wouldn't necessarily need to personally own autonomous vehicles. Yet, dealer networks' existing competencies of managing capital requirements and servicing cars could position them as key players operating and maintaining a network of autonomous vehicles in a region. Such a change would require a significant shift in strategy and competency from sales to operations.

2. **Auto manufacturers.** If the auto industry no longer sells primarily to individuals, the design options available to the industry also change, potentially shifting more toward utility than customer preference. The scenario raises the possibility for designing cars that are not optimized for passenger loads composed of individuals or family units and their cargo. We could see greater development of single passenger vehicles, larger vehicles capable of customized mass transportation, or smaller vehicles designated for cargo.

3. **Auto insurers.** For insurance companies like Liberty Mutual, auto insurance makes up a significant part of their business. Not only would autonomous vehicles shift the need for insurance (e.g., autonomous vehicles may have fewer accidents), but it would also shift who would need insurance (e.g., liability for accidents could shift to the makers of the technology). Most auto insurers, including Liberty Mutual, have recognized these emerging disruptions and begun to invest in innovation as well as differing operating models to prepare for what could be a radically different future market.

4. **Government.** The shift toward autonomous vehicles has implications for government services. Autonomous vehicles can collect data through their sensors, and this data can be uploaded to the cloud and sent back to the cars to optimize routing and traffic flows. It could challenge the dependence on public transportation. Why wait for a bus or train when you can have on-demand access to autonomous vehicles? Platforms could use data from demand patterns to combine rides along popular routes, which is the goal of Uber Express Pool. Different types of traffic control systems may be necessary in a world where cars can communicate directly with one another about routes, speeds, and intention, limiting the need for posted rules.

5. **Retail and restaurants.** Retail stores could begin to use autonomous vehicles as a delivery infrastructure. The restaurant-ordering company Olo has recently expanded its Dispatch platform, which integrates restaurants' software with Uber's to enable on-demand delivery drivers.[11] This software can also adjust delivery radius, food preparation schedules, and fees to optimize food quality through the delivery system. In this scenario, autonomous vehicles become available to all businesses, which has implications for store design and placement. Olo has recently partnered with Amazon to deliver these services.[12]

6. **Real estate.** Autonomous vehicles will also have implications for real estate valuations. Many urban locations may become more valuable in certain situations because parking would no longer be a constraint to access. Urban parking decks or lots can be converted to more valuable uses. Conversely, suburban locations may also become more valuable as traffic is reduced and people can use the commuting time for tasks other than driving. These shifts have implications for where companies choose to build office locations.

We could go on, but we hope we have made our point. The strategic implications of the adoption of autonomous vehicles will be significant and will span multiple industries. Will all these trends actually come true? No one really knows for sure, but it seems likely that at least some of them will, as well as others we haven't mentioned. Yet, this thought exercise raises the question: Has your company engaged in a similar thought exercise about how autonomous vehicles (or another technology) might reshape the competitive environment around your company's business model? This exercise simply represents a small set of general possible implications for a single technology. Autonomous vehicles may affect your industry differently than the possibilities mentioned here. Other technologies—such as 3-D printing, virtual and augmented reality, the Internet of Things, blockchain, and artificial intelligence, just to name some big categories—all have a similar sort of inevitability and may have a greater effect on your business. What are managers to do in the face of all this strategic uncertainty?

Reverse Engineering an Industry

Another productive approach is to think through how various technology trends are likely to reshape your industry. The advantages of this approach were apparent at a health care IT conference we attended recently. While the other presenters were thinking about how to drive the next step in the latest electronic medical records (EMR) adoption, we invited participants to conduct a similar thought experiment about the impact of information technology on the industry in the next ten years or so. It quickly became apparent that by the time they drove the EMR adoptions they were discussing across the industry, the health care industry would likely be entirely different. Unstructured data and artificial intelligence support may make the very EMR systems they were working on obsolete by the time they were finished. Participants also noted that large physician groups were essential because the IT infrastructure was so expensive. Blockchain technologies, however, could enable data to flow across organizations and enable continuity of care between organizations, potentially threatening the need for large physician networks to manage data and provide continuity of care.

The participants at the health care workshop—all smart, committed, sophisticated people—were making a mistake in strategic thinking by planning to adapt their organizations to the digital infrastructure as it exists today. They hadn't considered that by the time they completed the adaptation to today's environment, the digital environment to which they had adapted would be entirely different.

Any quarterback, soccer player, hockey player, or skeet shooter can tell you that you need to lead a moving target if you are going to hit it—aiming where it is going to be rather than where it is. Wayne Gretzky, the professional hockey player, is often quoted for his insightful comment, "I skate to where the puck is going to be, not where it has been." But, he also said, "You miss 100 percent of the shots you don't take." Like "The Great One," digitally maturing companies recognize technology as a moving target and begin to adapt their organization to the infrastructure of the future.

Takeaways for Chapter 4

What We Know	What You Can Do about It
• The existence and socialization of a clear and coherent digital strategy is the single most important determinant of a company's digital maturity. • Insufficient technical skills and lack of management understanding of digital trends are usually not the top barriers to the success of digital strategies. The more likely barrier is competition with other priorities for management attention and resources.	• Invest time in creating and documenting your digital strategy, with a focus on helping individuals understand how they would behave differently in the reality envisioned by the plan. Make sure that your documentation clearly identifies the "from-to" aspects of the strategy. • Give employees the opportunity to create and share their "future stories" (i.e., what it would be like if the reality envisioned by the strategy were achieved). • Engage your customers and ecosystem partners in pressure testing the strategy and stories. Make sure that your strategy helps them understand how their interactions with you will change as a result if the reality envisioned by the strategy is achieved. • Assess the key barriers to the success of the strategy and identify the specific actions required to address and overcome these barriers.

5 The Duct Tape Approach to Digital Strategy

Digital strategy is about adapting the organization to a changing environment in a way that leads to a sustainable competitive advantage. The academic concept of *affordances* can facilitate this proactive view of digital strategy. The concept of affordances was originally introduced in the psychology literature by James J. Gibson, referring to the possible ways that humans or other animals can interact with their environment. It treats the animal or the environment not as fundamentally separate from one another but as inextricably entangled. The environment determines the actions available to the animal, and the animal (particularly humans) can alter the environment in ways that change its capabilities for acting. For example, a key affordance of the electric lightbulb is the ability for people to see at night. That same affordance also restricts the potential actions of other animals, particularly nocturnal ones.

Gibson, a professor at Cornell University, spent World War II as the director of an Army Air Forces unit, where he studied the effect on visual perception of flying an aircraft. He used the term "affordances" in many works, including his 1966 book, *The Senses Considered as Perceptual Systems*. In his 1979 book, *The Ecological Approach to Visual Perception*, he offers the following definition of affordances—"The affordances of the environment are what it offers the animal, what it provides or furnishes, either for good or ill. The verb to afford is found in the dictionary, the noun affordance is not. I have made it up. I mean by it something that refers to both the environment and the animal in a way that no existing term does. It implies the complementarity of the animal and the environment."[1]

This concept of affordances was later picked up by the fields of computer science and information systems to describe how people interact with technology. Information technology changes how people and organizations can act in a particular environment, enabling new opportunities for action that would not be possible without the technology. Again, an affordance perspective suggests that people and information technology are fundamentally entangled with one another. Not only does technology change the possible actions of people and organizations, how people and organizations use technology changes the effects of the technology in practice.

More recently, this perspective has been extended to the organizational level. Technology can change the organizational environment in which they are used, enabling a new set of affordances. As companies learn and employ new digital affordances, the organizations will need to change in response. In fact, Harvard Business School professor Carliss Baldwin and coauthor Kim Clark boldly suggest that over the past century, organizations have been designed to serve the needs of the dominant technology of the age.[2] A key implication of an affordance perspective on digital strategy is that it shifts the focus from the features of the technology to how technology enables new strategic actions for people and organizations to engage.

Duct Tape and Digital Strategies

At the most basic level, an affordances perspective suggests that merely owning and implementing technology is not enough to deliver business advantage. While this insight may seem obvious at first glance, it is striking how often organizations behave as though it isn't. They either believe that the mere adoption of the latest technology will improve their business prospects, or they focus all their efforts on implementation, without applying the time or resources to make the types of organizational changes needed to benefit from the possibilities the technologies offer.

What may be the most paradigmatic example of a nondigital object enabling multiple actions is that of duct tape. Duct tape arrived on the scene in the 1940s, courtesy of scientists at a division of Johnson and

Johnson. Their objective was a durable and flexible tape that could be used in various wartime applications. The original version was army green and its original name, duck tape, reflected its water-repellant character. Post–World War II, the product found new uses in construction, including holding metal ducts together. A change in color to gray and a change in name to "duct tape" followed.[3] Even though the early civilian application for duct tape was to literally seal ducts, the possible actions this object enables in different settings is staggering. It can be used as a clothing decoration; to build wallets; to fix various issues in spaceflight; to stabilize helicopter rotors, as it was used during the Vietnam War; to craft vessels for carrying water; and even as a treatment for warts, among other uses (we confess that reading the Wikipedia article on duct tape, where we got most of these examples, was fascinating).

An Amazon search for "duct tape" reveals dozens of books describing potential uses for it in home improvement, crafting, medical uses, and team-building games, among others.[4] The first book listed is *A Kid's Guide to Awesome Duct Tape Projects: How to Make Your Own Wallets, Bags, Flowers, Hats, and Much, Much More!* One set of authors has a series of nine (!) books on uses for duct tape, and the television show Mythbusters has a series of episodes featuring the object. To say there is only one "right" use for duct tape is silly. It can be used in various ways, depending on your needs and interests.

Likewise, it is silly to say there is only one "right" way to use certain technologies. Twitter is one example of the digital equivalent of duct tape. Companies have learned to use this platform in surprising ways, depending on their needs. Some companies—like many of the major media outlets—use Twitter as a means of broadening the reach of their content. Others, such as Delta, JetBlue, and Southwest, use Twitter as an effective customer service tool, enabling them to support customers in a fluid service environment. Still others use Twitter as a business intelligence tool. The health care company Kaiser Permanente uses data generated by customers on Twitter to identify areas of improvement in business operations. After discovering that a lack of parking was the most common complaint from customers, the company used the geo-tag data to identify which facilities had the most significant issues. Auto manufacturer Nissan uses Twitter to help inform its marketing campaigns. The company livestreamed the presentation of its GT-R supercar using Twitter's video streaming app Periscope at a large national auto show. Real-time feedback from fans on Twitter revealed what aspects of the new car design were of most interest, information Nissan then used to design its marketing campaign.[5] The American Red Cross and the US Geological Survey use keyword monitoring to more quickly identify natural disasters than might otherwise be possible through traditional channels.

In the United States, we have seen Twitter emerge as a potent political tool as well. Political figures can share their thoughts and messages in real time, and armies of automated Twitter accounts can amplify certain messages to make the perspective seem more widely held than it is. While not all technologies necessarily enable the wide variety of strategic actions that Twitter does, the point here is simply that there is often more than one "right" way to use a particular technology to support business goals. And the challenge for organizations is to figure out in what ways a tool or platform might work for them.

Discovering Hidden Affordances

Another implication of an affordance view of technology in organizations is the recognition that some of the possible strategic implications of technology may not be immediately obvious. The affordance literature refers to these as "hidden" affordances, possibilities for acting that we may not immediately recognize. The potential strategic actions of these hidden affordances only become clear once an organization begins to use it and becomes more aware of its potential. An affordance perspective suggests that the path toward digital maturity is a recursive process in which technologies and the organizational environment mutually influence one another over time, rather than being a linear progression.

Technology creates new opportunities to work differently, and working differently creates new opportunities to infuse technology into the work process. The organization often needs time to figure out how to appropriate new technologies for business advantage and adapt its processes and structure to facilitate these new possible actions. For example, we interviewed one company that had adopted an expertise identification tool to help individuals in the organization locate and tap into experts. The tool analyzes digital content generated by employees and automatically generates knowledge profiles for them. Although the intention of the technology was to make others in the organization aware of what knowledge employees possess, the greater impact is in helping employees understand what knowledge they possess that is most valuable to others. This often differs considerably from their formal roles or how the employees think they are most valuable. The most valuable strategic advantage the company realizes from this technology was not even considered when the technology was initially implemented. Employees can also identify the hidden affordances encountered in their work. Once leaders learn of these new use cases, they can highlight them to the rest of the organization.

KLM's Evolving Use of Twitter

We return to the example of Twitter to illustrate the concept of hidden affordances further. The Dutch airline KLM evolved its use of Twitter in unexpected ways with its increased use of the technology. Like many airlines, KLM first used Twitter as a platform for social media marketing.

When the 2011 Iceland volcano eruption disrupted air travel across Europe for days, however, KLM realized that its Twitter presence could be a powerful customer service tool for communicating the company's plans for overcoming service disruptions. As a result, the airline began to use Twitter more heavily for customer service, because KLM leaders realized that it was an efficient and effective way to assist customers.

Once they began using it for customer service, they realized that it could also serve as a potent lost-and-found system, to return lost items to passengers after they had cleared security. Passengers aren't permitted to go back through security to retrieve the items, but employees can go out and deliver them to customers. As KLM used Twitter for one business purpose, other uses became apparent.

At the same time, these new action possibilities realized by continued use also identified organizational barriers to those uses. For example, as the airline relied more heavily on Twitter and other social media outlets as a customer service channel, KLM leaders realized that—as a global carrier—it needed to support many different languages across multiple time zones. The organization needed to develop further resources to adapt to the new actions the technology enabled to effectively execute them.

Progressive Affordances: Walk before You Run

The concept of hidden affordances suggests that organizations need to grow into effective digital strategy. In our survey, the goals of digital strategy differ by maturity stage. Early stage companies focus mostly on improving customer service and engagement. In addition to these goals, developing companies tend to focus more on improving innovation and business decision making. The maturing companies, however, are most likely to add transforming the business to these strategic goals. Indeed, at the highest levels of digital maturity, all these goals come into play.

These data lead to several possible implications. First, they may mean that companies need to learn to walk before they can run with technology. Early stage companies shouldn't try to jump straight to transforming their businesses if they haven't first learned the basics of improved customer service and efficiency. By focusing on these basics, companies can begin utilizing the technologies and make the necessary organizational changes to maximize their impact. This use will then begin to reveal hidden affordances, possibilities for innovation and decision making, as companies build on these more basic capabilities. Once companies have mastered new forms of innovation and data-driven decision making, they will be ready to begin thinking about transforming their businesses. The organization may not know enough to transform at early stages.

Second, our results suggest that the maturity levels we found may actually be differences in kind instead of incremental improvements. That is, maturing companies work in profoundly different ways than do early and developing companies. To progress to the next stage, becoming better at digital efforts may be insufficient; companies may need to approach those efforts differently to continue to the next stage of maturity. What got you to your current stage may not be enough to get you to the next one.

Third, our data may also mean that the lead of digitally maturing companies over less mature companies will begin to compound. The most mature companies are still trying to do things differently to leverage their advantages. The real path toward digital transformation may only begin once companies reach the maturing stage. In martial arts, a black belt is a symbol not of expertise but of maturity, in which the student has mastered the basics of the art. At that point, "you have learned the basics, and now you are ready to start your real training."[6]

We've often speculated about a hypothetical fourth category of maturity, which maturing companies are striving to attain but will never fully realize. Yet, the goal of attaining this Platonic ideal of a digitally maturing company is enough to keep the organization striving to become better and to do business differently. We address this topic in more depth in chapter 15.

False Affordances: Digital "Placebo Buttons"

An affordance view of digital strategy also raises the possibility of false affordances. False affordances are actions that do not have any real function. A good example of a false affordance is a so-called placebo button. A recent *Boston Globe* story reported that the buttons pedestrians can press at crosswalks to get a "walk" signal don't actually do anything.[7] Yet, people may be more willing to wait for the signal if they feel their response has been recorded. Likewise, people often push an elevator button multiple times if they are in a hurry, even though they know it does not speed up the elevator, and, in most cases, the "close door" button has no effect on closing elevator doors.[8] Digital strategies are particularly susceptible to false affordances. False digital affordances are those that may give the perception of an organization being more digitally mature without actually contributing to the effective functioning of the organization. Managers are often attracted to flashy tools that provide a "wow" factor, without considering whether they will actually change how the organization does business—the digital equivalent of the "walk" or "close door" buttons.

Nowhere is this point clearer than a comparison between two clothing brands.[9] One upscale clothing retailer invested in flashy technologies in its flagship store in Manhattan, such as digital dressing rooms, RFID-tagged clothing, and sophisticated software that allowed the systems to make recommendations for products that went with the items the customer had selected. This technology was expensive and turned out to be more of a novelty than a delivery of real value to customers. The technology ended up getting shelved when it turned out to be more trouble than it was worth.

In contrast, the retailer Zara—one of the fastest growing companies in the world[10]—is often credited with using an IT infrastructure to deliver "fast fashion" by looking for inspiration from popular designs and getting the clothing to the marketplace in a matter of days. Yet, the IT infrastructure that drives these advantages is relatively rudimentary from a technological perspective. Zara spends about 25 percent of the industry average on technology, yet Zara has continued to grow, despite struggles in other parts of the retail industry. This comparison makes clear that the acquisition and deployment of advanced technology is not a substitute for a strong digital strategy. Technology is secondary to the strategy it enables.

Indeed, digitally maturing companies think about strategy differently than their less mature counterparts (figure 5.1). Respondents from early stage companies reported that their organization talks more about digital business than does anything about it. Respondents from developing-stage companies noted that their companies think of digital strategy as one-off initiatives that do little to affect the core business. In contrast, respondents from digitally maturing companies reported decisively that their company's digital strategies are a core part of the organization's business strategy.

This response continues a theme that we've seen manifest in our research across the years of the study: Digital strategy isn't just thinking of new initiatives that enable organizations to do business in the same way but slightly more efficiently. Instead, it involves fundamentally rethinking how you do business in light of all the digital trends occurring both inside and outside your organization. It involves identifying potential new services, sources of revenue, and ways of interacting with employees.

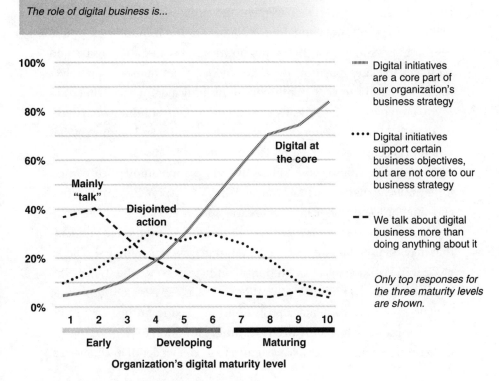

The role of digital business is...

Figure 5.1

Collective Affordances

Much of the affordance literature addresses the relationship between the individual and the environment. Organizations won't likely work effectively if all the stakeholders are using the technology in significantly different ways. UC Santa Barbara professor Paul M. Leonardi introduces the concept of collective affordances to recognize the need for an organization to use the possible actions performed by technology in a way consistent with or complementary to other users within the organization.[11] Groups that tended to gravitate toward a common set of actions with a new technology performed better than groups in which individuals used the technology in divergent ways.[12]

The need for collective affordances raises the importance of strong communication between management and employees to effectively enact digital strategy. It is not enough to implement a new technology. Employees must also know what they are to do with it, because the possible and most valuable uses may not always be obvious. Communication involves leadership clearly and coherently communicating a vision for digital strategy to employees. It also means listening to input from those employees on how that vision is being executed in the real world and modifying that vision to account for actual ways technology is being used.

Employees can also serve as effective sensors for unintended barriers to or implications of a digital strategy. They can report these difficulties, allowing the leadership team to adapt the strategy accordingly. Nonexecutive leaders can be valuable first responders to tackle unexpected problems that may occur while executing a digital strategy. As such, they should be given some latitude to act when they identify issues once the strategy has been effectively communicated to them. In fact, our research shows that digitally maturing companies push decision making further down into the organizational hierarchy to enable the organization to respond to digital trends more quickly, a topic we return to in greater depth in chapter 12.

Takeaways for Chapter 5

What We Know	What You Can Do about It
• An **affordance perspective** suggests that the value of technology is found in the new capabilities it enables for your business, not simply in owning it. As with duct tape, a single technology may elicit many possible strategic moves. • **Hidden affordances** refers to strategic moves enabled by technology that aren't apparent when you first adopt it. They reveal themselves as you begin using the technology. • **Progressive affordances** suggest that certain capabilities must be mastered before subsequent capabilities can be tackled. Companies often move from efficiency/customer experience to improving innovation/decision making to business transformation.	• Develop processes for sharing by using case scenarios for technology across your organization, particularly uses of technology that are not expected or anticipated. Help the organization learn from the success or failures of others' efforts. Make this a standing component of management meetings and communications. • This institutional sharing of lessons learned should be a critical component of the learning pilots identified in the Takeaways for Chapter 2.

II   Rethinking Leadership and Talent for a Digital Age

Leadership. The word "leadership" generates powerful images for most of us. We associate the notion with all sorts of real and fictional figures, ranging from warriors, government officials, and clergy to political activists and CEOs. Each of us has our own personal vision of the leader that we are ready to follow whenever, wherever. Sometimes this vision is based on fact or personal experience. Often it is based on the mythology that has sprung up around an individual and is embellished over time. If you go to Amazon, you will find hundreds of tomes on individual leaders and the subject of leadership. Many of these books purport to share previously undiscovered secrets of leadership. Examples of this genre range from *The Leadership Secrets of Attila the Hun* to *The Leadership Secrets of Santa Claus: How to Get Big Things Done in Your "Workshop" . . . All Year Long.* Alexander Hamilton, the subject of a recent hit Broadway musical, merits an entry in this genre with an audiobook entitled *The Leadership Secrets of Hamilton: 7 Steps to Revolutionary Leadership from Alexander Hamilton and the Founding Fathers.*

Our intense interest in leadership is not surprising. In times of disruption and radical change, we yearn for leaders who will help us persevere and succeed. As organizations seek their footing in a turbulent business environment, they need outstanding leaders to guide the organization to a new reality. Leaders not only need to create a vision that people can rally around; they also need to create the conditions that enable digital maturity, largely by attracting the best talent and bringing out the best in the talent they attract. But we desperately want to know whether we

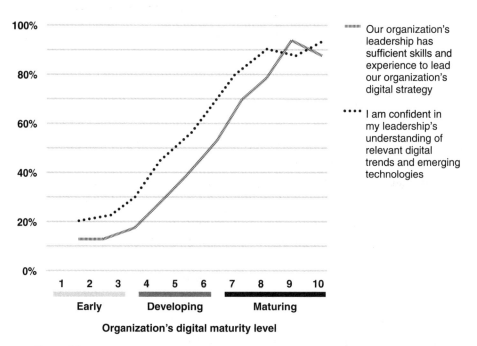

Our organization's leadership has sufficient skills and experience to lead our organization's digital strategy

I am confident in my leadership's understanding of relevant digital trends and emerging technologies

Organization's digital maturity level

Figure 6.1

need a Henry Ford or a Mahatma Gandhi or a Winston Churchill or a Phil Jackson or a . . . pick your favorite leader.

Our research finds that effective leadership is one of the most critical factors associated with digital maturity. Between 20 and 30 percent of respondents from early stage companies expressed confidence that the leaders of their organizations have the skills and experience to direct digital strategy (figure 6.1). By contrast, the confidence in leadership is approximately 90 percent among respondents from maturing companies. The inevitable question is, What can early stage companies learn from digitally maturing companies about the characteristics of effective leadership?

The Challenge of Digital Leadership

The rapid changes associated with digital disruption c.. ing, challenging our notions about the nature of leadership. Mai.y us fervently believe that different times call for different types of leaders. In times of war, a country needs a different leader than in times of peace. In times of prosperity, a country needs a different leader than in times of economic crisis. Need an example? See the *Fortune* blog entry by Gautam Mukunda, an assistant professor at Harvard Business School, entitled "Jefferson and Lincoln: Different Leaders for Different Times."[1] Our deeply held beliefs in "different leaders for different times" color any discussion about leadership in a digital environment. Some argue, "All the rules have changed in the digital age. The leadership handbook needs to be completely rewritten." Many jump to the conclusion that leaders who were once effective no longer can be.

But, is this true? Does the essence of leadership really change in the digital age? Do we really need to press the reset button? Or are greater and greater levels of uncertainty causing us to forget the essentials and focus on the latest bright and shiny object? Is it possible that more is the same than is different, but we're focusing on what's different because we're so alarmed by the threats to the status quo? Are our views on effective digital leadership complicated by the fact that many individuals responsible for leading in a digital age have roles like chief marketing officer and chief talent officer, which have historically not had to deal extensively with rapidly changing technologies?

Let's assume, for a moment, that some characteristics of effective leadership are unchanging, whereas others are more closely tied to the particular demands of digital disruption. The question is which are which? Which principles of effective leadership transcend digital disruption, and which principles require adaptation? When is it appropriate to stick with what has worked, and when is it appropriate to update one's leadership playbook?

Leadership for a World of Scalable Learning

In the digital age, when the pace of change is accelerated and much is new and unknown, learning is critical. John Hagel, cochairman at Deloitte's Center for the Edge, has been studying frontline workgroups at various organizations and describes the hope they bring.[2] These are small groups, from three to fifteen people, who spend the bulk of their time together doing work that is intertwined and can't be pulled apart. They range from groups of maintenance workers to hospital emergency wards, and they were largely formed bottom up, improvising on the fly and inventing as they went. More than other working groups, these teams are able to learn faster and accelerate performance improvements, creating new knowledge along the way. But it takes a certain type of leadership to enable this environment to thrive and to propel these groups forward.

Traditionally, organizations (and leaders) have been focused on scalable efficiency. Hagel describes the mark of a strong leader in that kind of environment: "The model of effective leadership in the past was somebody who had all the answers, who knew exactly the direction we need to go in and what needs to be done and can tell me what to do. No matter what the question, you can count on that leader to have the answer. And God forbid if they say they don't have an answer[;] it's a sign of weakness. Fire that person. Get somebody else in who does have the answers."

In a digital environment, organizations must shift from a world of "scalable efficiency" to one of "scalable learning." This means leadership needs to shift, too. As Hagel describes, "Leadership in the future is not about having all the answers; it's actually around being able to frame the right questions—powerful and inspiring questions—where the leader is saying, 'I have no clue, but this is a really important question. And if we could figure this out, we would do amazing things.'" That's a different model of leadership—one focused on creating an environment that inspires people to learn faster and moves the group to work together to find the answers.

Wilhelm Johannsen: Genotype vs. Phenotype

Evolutionary science provides a useful way to frame our answers to the question about which characteristics of effective leadership are enduring and which are the ones that require adaptation to a digital environment. In the world of evolutionary science, the name Charles Darwin looms large. Concepts like natural selection and survival of the fittest were first introduced in Darwin's famous 1859 book *On the Origin of*

Species. Far fewer people, however, are aware of the name W̶ Johannsen. Johannsen, a Danish botanist, conducted a series of p studies in the late nineteenth and early twentieth centuries. He dete mined that he could produce large or small plants from seeds that carried identical genes. Johannsen introduced the terms "genotype" and "phenotype" to explain this phenomenon in a landmark essay (and follow-up book) that became one of the founding texts of genetics.

Johannsen used the term "genotype" to describe the set of genes that an organism carries. The genotype is the genetic blueprint for the organism; it is fixed at conception and remains unchanged through the organism's life. The phenotype, by contrast, describes the physical characteristics of the organism, which results from interactions between its genetic blueprint and the environment. The same genotype can produce very different phenotypic characteristics, depending on an organism's environment. For example, although a person may carry genes for being tall, height is also influenced by diet, climate, illness, and stress, among other environmental factors.[3]

This metaphor is an apt one for understanding the nature of digital leadership. Note that we are *not* referring to genes of individuals but rather the traits of a good leader. Consistent with the themes of this book, we believe strong leadership is a trait that can be learned, not one an individual is born with or possesses innately. The genotype, or blueprint, for effective leadership consists of a series of characteristics, ranging from providing purpose to inspiring employees to facilitating collaboration. These traits, which have always been part and parcel of good leadership, will continue to define the blueprint for good leadership. Yet, those fundamental traits will be expressed differently in a digital environment than a more traditional one.

Fundamentals of Leadership

Hundreds of books have been written on leadership, from Stephen Covey's *7 Habits of Highly Effective People* to John C. Maxwell's *21 Indispensable Qualities of a Leader*. We won't attempt to substitute for

nd resources that are already available. But we
lership capabilities that are particularly impor-
re discuss in greater detail in this chapter and

vision *and purpose.* Having vision and setting
always been a leadership fundamental. But in a digital
ronment, this capability takes on new meaning. In our study, it's
cited as the most important leadership skill. With the uncertainty
and undiscovered possibilities facing today's business environment,
leaders must have a transformative vision for their organizations,
with both long-term and short-term outlooks.

- **Business judgment:** *Making decisions in an uncertain context.* Leaders
  have always had to demonstrate commercial acumen and wisdom
  through sound business decisions. In a more digital environment,
  decisions often must be made more quickly and with incomplete or
  uncertain information. Leaders may no longer be able to wait for an
  ROI analysis or early performance measures.

- **Execution:** *Empowering people to think differently.* Leaders have always
  had to rely on others to achieve results. To do this effectively in a
  digital environment, leaders must empower their people to think
  more creatively, work more collaboratively, adopt a more entrepreneurial mindset, and be leaders themselves.

- **Inspirational leadership:** *Getting people to follow you.* Digital transformation and digital maturity are huge changes. To navigate people
  through the change and uncertainty and get behind a vision, leaders
  must inspire others. People must *want* to follow them, not be forced
  to do so.

- **Innovation:** *Creating the conditions for people to experiment.* Innovation is critical for organizations that want to compete. But innovation requires experimentation, continual learning, and—most
  important—risk taking. Yet, as we discuss in later chapters, fear of
  failure is a huge impediment to experimentation and innovation in
  many organizations. Leaders must overcome this obstacle and create
  an environment that encourages genuine experimentation.

- **Talent building:** *Supporting continuous self-development.* Developing people has always been a core leadership trait. In an environment where continuous learning is critical to building capabilities, leaders must empower and enable self-development. This includes providing employees with opportunities to take on new challenges, as well as supporting self-directed learning outside the organization (see chapters 8 and 9).

- **Influence:** *Persuading and influencing stakeholders.* As we discuss in later chapters, hierarchical structures with solid lines on org charts will diminish. Instead, organizations will look more and more like peer-to-peer networks. Power through simple authority rarely exists. Instead, influence and persuasion become key to building support and getting things done.

- **Collaboration:** *Getting people to collaborate across boundaries.* As we discuss in chapter 13, collaboration is a core trait of digitally maturing organizations. Leaders must encourage and enable this collaboration internally by breaking down silos and increasing cross-functional teaming, as well as outside the organization by extending partnerships and blurring the boundaries of the organization.

Diagnosing Mistakes in Understanding the Nature of Digital Leadership

The genetic metaphor of genotype and phenotype help us articulate three mistakes that leaders often make when it comes to digital leadership.

1. First, as we previously discussed, many leaders mistakenly believe that the genotype for successful leadership fundamentally changes in digital environments, that somehow the core of good leadership is dramatically different because of digital challenges and capabilities. This mistake can cause leaders to ignore many tried and true leadership lessons and experience in an attempt to do things entirely differently. Good leaders can make bad decisions if they ignore valid leadership instincts honed over a career simply because they were shaped in a different environment.

2. The second mistake is the converse of the first, when leaders think that the phenotype of good leadership will somehow be unchanged in a digital environment. While good leadership remains good leadership, it will necessarily be expressed differently in a radically new environment. Just like we don't see carbon paper, typewriters, and adding machines in today's offices, leaders need to update approaches to leadership that may have been forged in the dot-com era.

3. Third, many leaders mistake the outward expression of digital leadership for the phenotypic expression of good leadership in a digital environment. Doing things digitally does not automatically make one an effective leader. For example, although effective communication in a digital environment likely involves using digital platforms, simply using these platforms does not automatically result in good communication. In fact, since these platforms can facilitate the transfer of all sorts of information, they can amplify bad leadership just as easily as good leadership.

What Stays the Same?

Let us turn now to our data, which identify key management characteristics that remain unchanged because of digital disruption, as well as those that take on new importance. This point is another one in which, when speaking to executive audiences about our research, people often ask, "How is this different from the way we have always done things?" Our response is often that it is not, but it's surprising that leaders often lose sight of the basics of good leadership in the face of digital disruption. Seeing how many respondents at companies with unsuccessful initiatives report these problems suggests that these key lessons bear repeating.

Digital Leaders Focus on the Business Value of Initiatives and Invest Appropriately

One important leadership characteristic that has not changed because of digital disruption is the importance of focusing on the value case for digital initiatives. Although this lesson may seem obvious, our

CONF. POINT!

data show that leaders often become so focused on the technological aspects that they forget why they are engaged in these efforts in the first place—to improve the way their company does business. Digital transformation is only partly about technology; it is also, and more importantly, about using new technology to enable novel or more effective business strategies. Leaders often believe that they need to be in mobile, analytics, artificial intelligence (AI), or other emerging technologies without being able to clearly articulate _why_ they need to invest in these technologies or what business purpose they could serve. Understanding the value case does not preclude experimentation with initiatives to determine their possible value to the organization, the subject of chapter 14. Indeed, research and development are critical enablers for success in a digital environment. Just be sure that you know why you are beginning an initiative and what your business goals are for it— even if it is intended to be a learning pilot.

LEARN IS A GOAL !

As a corollary to this rule, don't forget to provide sufficient investment for your initiatives to succeed. It's surprising just how often leaders expect a project to go well without giving it proper financial support and resourcing. We found clear evidence for this tendency in our research. We asked respondents whether their digital initiatives were generally successful or unsuccessful (figure 6.2). Perhaps unsurprisingly, we found that appropriate levels of investment are a key factor in success. When respondents reported that their initiatives had the right level of investment, 75 percent of them said that those initiatives were successful. In contrast, when respondents said their company did not commit sufficient time, energy, and resources, only 34 percent reported success.

Investment also involves investing in the right things. Often leaders think that once the system is implemented and in place, the investment is complete. On the contrary, management scholars have long recognized that introducing technology changes how people using those tools work together.[4] If digital transformation is essentially an organizational and people issue, as we argue in this book, the real investment in digital transformation is only part technology. It will take time for

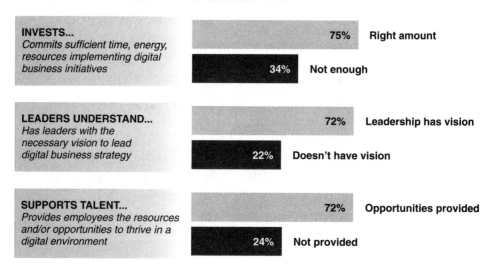

Percentage of respondents who say their organization's digital initiatives are successful when their organization...

INVESTS...
Commits sufficient time, energy, resources implementing digital business initiatives

75% **Right amount**

34% **Not enough**

LEADERS UNDERSTAND...
Has leaders with the necessary vision to lead digital business strategy

72% **Leadership has vision**

22% **Doesn't have vision**

SUPPORTS TALENT...
Provides employees the resources and/or opportunities to thrive in a digital environment

72% **Opportunities provided**

24% **Not provided**

Figure 6.2

people to learn to use the new technology and for the organization to adapt its work and communications processes to accommodate it.

Somewhat disturbingly, our data suggest that the gaps between those companies and leaders who are investing appropriately and those who are not may continue to widen (figure 6.3). Respondents whose companies were not currently investing enough in digital initiatives did not see an increase in investment in the near future. In contrast, those who are already investing enough are far more likely to be increasing that investment.

Digital Leaders Lead from the Front

Top management support is also key to digital maturity. Leaders who aren't directly involved with the sourcing or creation of new technology often assume that they are not "digital" leaders. But as companies begin to engage more heavily in digital business, all leaders must become digital leaders. Whether directly involved in implementing

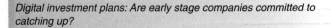

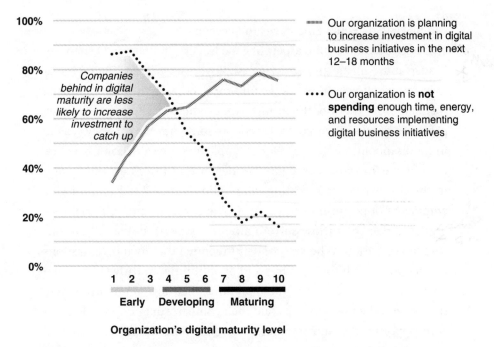

Figure 6.3

the technology or not, leaders must understand the value case for digital initiatives and what other aspects of the organization need to be aligned to accomplish those goals.

When executives simply delegate responsibility for digital business to the technologists, it is a recipe for near-certain failure. Not only do top management involvement and direct support for digital business initiatives signal to the company that those initiatives are important, this support can help align other aspects of the organization with these goals. Of respondents from organizations with leaders who had the necessary vision to lead digital strategy, 72 percent reported successful initiatives, while only 22 percent of respondents with leaders who lacked vision reported success in their initiatives. The good news is that it's often much easier to teach executives what they need to know about digital business than it is to equip technologists with the leadership experience and strategic insight they would need to lead digital business efforts effectively.

Digital Leaders Equip Employees to Succeed

This brings us to the third aspect of good leadership that hasn't changed. Leaders must enable and empower employees to succeed. Digital initiatives cannot, however, be successful simply because they have a strong mandate from the top. If you just expect employees to engage in new processes because your company adopts some new technology, you're in for disappointment. They won't. Employees typically don't have the time or the know-how to figure out new ways to work on the fly and in the context of their existing job responsibilities. Leaders must give employees opportunities to succeed. The contrast among these findings is starker than those surrounding investment, but similar to our findings for vision. When respondents reported that their organizations provide them with the resources and opportunities to thrive in a digital environment, 72 percent of them said that their digital initiatives were successful. When respondents said their company did not provide them with opportunities and resources, however, only 24 percent reported successful digital initiatives.

These opportunities can come in many forms. Employees should be provided with adequate training to learn to engage the technology and associated processes effectively. Training need not take the form of traditional classes; it may simply mean ensuring that adequate resources are available online to help them learn (and ensuring that employees are aware of them). Alternatively, it may mean that employees are moved within the organization more frequently so that they can learn other ways of doing things from coworkers. Employees must be given time and space to adapt. They are likely to be good at sticking with established ways of doing things that are safe and familiar. New ways of working require adequate time and cognitive resources to explore and learn.

Digital Leadership Is Not Magic

The science-fiction writer Arthur C. Clarke notes that any sufficiently advanced technology is indistinguishable from magic.[5] The same is true of technology. Many leaders view technology as a kind of sophisticated magic (or as a con job) simply because they don't understand the basic principles behind it. When we peer behind the curtain to see how the magic is done, however, we see it's just some people pulling the same old knobs and levers that leaders have always pulled. Yes, those knobs and levers may look a little bit strange, and the effects of pulling them may be unfamiliar, but digital leadership is not fundamentally different. Digital leadership is just leadership, albeit in a somewhat new environment. These principles are not magic, nor are they overly difficult to understand at a sufficient level to be an effective leader.

Takeaways for Chapter 6

What We Know	What You Can Do about It
• Strong leaders are critical for digitally maturing companies. Of respondents from digitally maturing companies, 90 percent said their leaders have the skills necessary to lead, whereas only 25 percent from early stage companies do.	• Starting with the principles outlined in this book (e.g., leading networks of dynamic teams, rather than traditional structural hierarchies), create your list of the most critical capabilities for effective leadership in a digitally maturing environment.
• There are core leadership skills that good leaders always need, but additional leadership skills are needed in a digitally maturing environment (e.g., managing networks).	• Assess yourself (and ask your colleagues to do the same), with the goal of identifying the most critical shortfalls. Conduct 360-degree assessments as well.
	• Ask each leader to create a game plan to address these shortfalls. Incorporate these plans into goal-setting and evaluation processes.
	• Triage existing leaders, with the goal of determining whether they have (or can develop) these capabilities or whether new leadership is required.

7   What Makes Digital Leadership Different?

Of course, just because much remains the same in good digital leadership doesn't mean that nothing changes. As the organizational environment shifts, the skills required to be an effective leader will likely change somewhat as well. How these characteristics manifest themselves in a digital environment, however, may vary from how they manifested themselves in previous environments.

We know, for example, that effective leaders are inspirational, with exceptional communication skills. Yet, the communication tools and platforms that an effective leader uses in the current environment vary considerably from the tools used just a few years ago. Today, we live in the world of Slack, Workplace by Facebook, Yammer, and Jive. In this environment, what used to pass as a swift response to a query or a controversy ten years ago will be viewed as painfully slow. Effective communicators use the best tools available to them for the right communication purpose. Managers likely even need to rethink how they use email to be effective communicators in today's environment.[1]

Another example? Trust. We know that establishing trust with employees, customers, and other stakeholders is another critical characteristic of good leadership. But the process by which a leader establishes trust in a digital environment filled with expectations for radical transparency will be quite different from the process used in a previous era. Not so long ago, organizations could reasonably control the flow of certain types of information, dealing with certain issues publicly and others in private. Sound advice from a wide variety of figures, such as the

Roman writer Publilius Syrus, Catherine the Great, and Vince Lombardi, suggest that leaders should "praise in public and criticize in private." Yet, in the current information environment, leaders cannot readily assume that any information can be kept private and must be prepared to deal with all situations publicly. As a result, managers must operate differently to engender the same types of trust that they could previously.[2]

Digital Leadership Is in Demand

Whatever digital leadership consists of, one thing that is clear from our research is that it is very much in demand. When we asked respondents whether their organization needs to find new leaders for the organization to succeed in the digital age (figure 7.1), 68 percent agreed that their

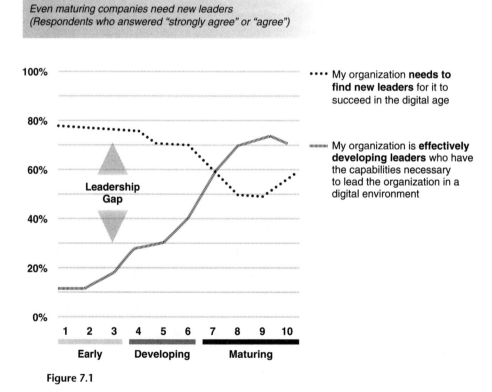

Figure 7.1

organization did, in fact, need new leadership to compete. What is more striking is the surprisingly little variance across maturity categories in these responses. While 77 percent of respondents from early stage companies reported needing new leaders, 55 percent of respondents from maturing companies said the same. In other words, respondents from more than half of the digitally maturing companies still said their leadership is lacking. Rather than lamenting the dearth of genuine digital leadership at most organizations, a better response is to recognize that digital leadership is truly a difficult challenge. As the environment continually changes around them, leaders must continually face new challenges and adapt both the organization and their leadership style to these new environments. Successfully meeting these challenges requires new skills and capabilities that leaders have not needed in the past.

A Leadership Perspective from Australia and New Zealand Banking Group

At the Australia and New Zealand Banking Group (ANZ), Maile Carnegie, group executive, digital banking, recognizes that leadership looks different in the twenty-first century. Her group has had to launch a fundamental recap of their definition of good leadership, recognizing that the leadership skills needed to run a hierarchical command-and-control organization are quite different from those needed to run a more agile or more distributed leadership organization. It involves two factors:

1. **Technical mastery:** She says, "The two big skills we are looking for are technical mastery and leadership. For the technical mastery, we need more software engineers. We need people who have still got their hands on the tools, have a craft, and are excellent at their craft. Unfortunately, in most twentieth-century companies, success has been defined by actually no longer doing the work but managing others who do it. You then have people who were at one point a great marketer or a great data scientist or a great software engineer. They then got promoted, and the promotion meant they were no longer actually doing their craft. So, the first thing to focus on is to get people back at being excellent at their craft—their technical mastery."

2. **Leadership:** "The second skill needed to be a great leader in today's context is to be inherently curious. Today's leaders need to lead through influence rather than through command and control. That's quite hard for people who have really only had one quiver in their leadership bow, which is command control."

(continued)

A Leadership Perspective from Australia and New Zealand Banking Group (continued)

> Developing these types of leaders at ANZ, Carnegie says, starts in the recruiting process, in which all roles were open, and all employees had to reapply for their job. And for leadership roles, the first criteria they screened for were culture and leadership potential. As she describes, "Not everybody had all the attributes we were looking for in our new agile world, but we believed that if they had the right cultural values and leadership attitude, they could develop the other skills relatively quickly."

There simply just aren't enough effective digital leaders to go around who can meet the challenges most companies are facing. But here is where the digitally maturing companies are distinguishing themselves from their less mature counterparts. Digitally maturing companies are doing something about the lack of effective leadership. When we asked whether the respondents' organizations are effectively developing the types of leaders who have the capabilities necessary to lead companies in a digital environment, the gap in responses was considerable. While 63 percent of respondents from maturing companies said that they are effectively developing the type of leaders they report needing, only 33 percent of developing-stage companies and a paltry 13 percent (!) of early stage companies said the same. While all types of companies are facing a lack of qualified leaders who are effective amid digital disruption, the maturing companies are far more likely doing something about it and actively developing those leaders.

Developing Digital Leaders at a Fortune 100 Company

> The chief digital officer at a Fortune 100 company speaks about how he develops the necessary digital leadership in his group. First, he focuses mostly on developing talent internally. He says, "A lot of our success is on actually growing our own talent. I've put 100 percent of the directors and up in my team, over fifty people, through training that's all about their cross-functional leadership skills and helping them become the best leaders they could be. This is multiple

days of training, coaching afterwards, assessment up front, very involved, very rigorous."

Furthermore, many of the skills he is seeking to develop in his leaders are not necessarily those that one might initially think of when thinking of digital leaders. His talent development model is not about providing leaders with digital skills, but about equipping leaders to operate effectively in an environment that is dominated by those tools. He says, "The kinds of leadership skills are about how you engage and inspire others. How do you create shared vision? How do you manage and deliver on programs cross-functionally? How do you resolve conflict when the priority of the objectives aren't aligned? Those kinds of integrated, or we actually call it interactive, leadership skills are incredibly important in the work we're doing. Frankly, if you listen to it, I didn't say anything about digital skills, actually. It's all about integrated, interactive leadership."

Skills for Digital Leaders

To understand which skills and capabilities are more important in a digital environment, we asked respondents directly. In one survey, we asked, "What is the most important skill organizational leaders should have to succeed in a digital workplace?" We then provided a simple text-box in which respondents could give a free-form response. Out of 3,700 respondents, 3,300 answered the question. Some responses were just a few words, while others were several sentences. Our research team categorized all 3,300 responses into similar groups (figure 7.2).

While the responses suggest that leaders need to understand technology, technical skills per se are not a prerequisite for effective digital leadership. Digital leadership is about leading amid the new business environment created by digital disruption, but it is not about mastery of technology. Characteristics such as being change oriented or having a transformative vision are reported as more important than mastery. Further, a closer examination of these responses reveals that they build on one another. Taken together, they paint a compelling composite picture of what effective leadership looks like in a digital environment.

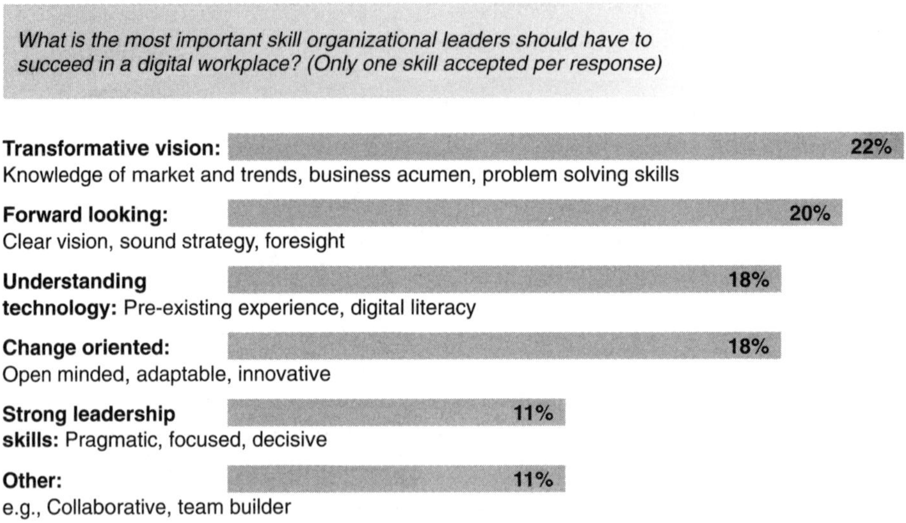

What is the most important skill organizational leaders should have to succeed in a digital workplace? (Only one skill accepted per response)

Transformative vision: 22%
Knowledge of market and trends, business acumen, problem solving skills

Forward looking: 20%
Clear vision, sound strategy, foresight

Understanding   18%
technology: Pre-existing experience, digital literacy

Change oriented: 18%
Open minded, adaptable, innovative

Strong leadership   11%
skills: Pragmatic, focused, decisive

Other: 11%
e.g., Collaborative, team builder

Figure 7.2

Transformative Vision and Forward-Looking Perspective

Providing vision and direction have been long-standing essential components of leadership. But in a digital environment, they take on new importance with an emphasis on future change. In our surveys, the most important skill that respondents reported is developing a transformative vision (22 percent), which includes knowledge of markets and trends, business acumen, and being a good problem solver. The second most important is being forward looking (20 percent), which includes clear vision, sound strategy, and foresight. These skill sets are clearly related to each other. We interpret the latter as an understanding of how business trends are evolving because of technology and the former as an ability to guide the business in response to those trends. Vision is about providing purpose and direction for the changes that are needed. This transformative vision is essential given the magnitude of the change required. John Glaser, senior vice president of population health at Cerner, provides a way to begin thinking of vision in an uncertain future. He says, "Work on things that are likely to be relevant to many possible futures. Tell me a future in which engaging patients to manage their own health is a bad idea, because I don't see that future at all. So, I may not know how it's going to play out, but under almost any conceivable circumstance, these things will be relevant."

Digital Literacy

AWARENESS CONCEPT

Understanding technology is listed as only the third most important skill. We need to examine the nature of these responses more closely to understand what this trait entails. Respondents said that leaders needed to have pre-existing experience and general digital literacy, as opposed to any hardcore technical skills like programming or data science. In other words, it is important for leaders to understand the general principles about how technology works, as well as the range of capabilities (and possibilities) that come with this technology.

Digital literacy among leaders is valuable in two ways. First, it is critical to supporting the first two leadership skills cited: having vision and being forward looking. A leader who is not digitally literate will not be able to keep abreast of emerging trends and developments to understand how those trends can bring new value to the organization and employees. For example, one executive recently noted, "How could we possibly have known 10 years ago that streaming services such as Netflix would become as dominant as they are?" Yet, we were teaching those very concepts in our undergraduate business school classes at around the same time. The strategic failure of this CEO was the lack of digital literacy that would have allowed him to anticipate emerging trends.

Second, understanding the general principles about how technology does (and does not) work enables leaders to make better, more informed decisions—all the more important in an uncertain environment. The good news is that this basic working knowledge is not difficult to come by, and it is often much easier and more effective to help established business leaders become digitally literate than it is to teach technologists the strategic knowledge they need to lead effectively. Digital literacy means that a leader can, among other things, recognize if a particular technology is or is not appropriate for certain business applications. The lack of digital literacy among governmental leaders was apparent in Mark Zuckerberg's 2018 testimony before Congress. The elected officials clearly lacked basic working knowledge of Facebook's business model and value proposition. Although the news media derided Congress's lack of knowledge, we suspect that it reflects the digital literacy level of many corporate board members.

This trend is particularly important, because our research shows that a leader's digital literacy is also associated with an organization's digital maturity. Only around 10–15 percent of respondents from early stage companies said they are confident that their leadership understands digital trends, whereas around 80–90 percent of respondents from maturing companies reported so. Kristin Darby, of Cancer Treatment Centers of America, echoes this sentiment: "If they have the dynamic skill sets to be a visionary, understand how mobile and digital can change experiences, and they can resonate with patients, I think that that's really what we're looking for, which is very difficult to find. But the technology will come along to support those skill sets. It's really that cultural fit and consistent vision and aptitude for technology."

If leaders have a strong understanding of the opportunities and threats technology represents, they are more likely to make the necessary organizational changes to bring about digital maturity. Of course, we would be careful to point out that this correlation does not imply causation. Perhaps digitally maturing organizations actively seek out digitally literate leaders, rather than digitally literate leaders creating digitally maturing organizations.

Digital literacy is essential because it is likely a necessary condition for the other important skills identified by respondents. It is difficult for leaders to have a transformative vision if they don't have any understanding of how or why the environment will transform and the capabilities the organization possesses to respond. They cannot be forward looking if they do not have any sense of what that future might look like, and they cannot have a transformative vision if they do not possess a working knowledge of the tools that will enable that transformation.

Change Oriented

Tied for the third most important characteristic, cited as first by 18 percent of our respondents, is that a leader must be change oriented—open minded, adaptable, and innovative. This response also provides an important qualifier for the earlier skill sets reported as valuable. Open mindedness is essential because leaders must be comfortable adapting to a fluid environment and be prepared to change course if the technology and market environments evolve in unanticipated ways.

This change-oriented mindset also applies to the knowledge that a digital leader must possess. A digital leader's knowledge "stores" must be continually updated to account for changes in technology. If your organization does not have a process in place by which your leaders can update their knowledge regularly, the digital knowledge of your leadership will gradually grow obsolete. Leaders can replenish their knowledge stores through various practices, such as formal continuing education, in-house education, cross-generational reverse mentoring programs, or any of an abundance of online programs.

Executing through Strong Leadership Skills

This leads us to the final leadership trait that respondents indicate is important—strong leadership skills. This category, focused on the driving and execution aspects of leadership, includes such characteristics as pragmatic, focused, and decisive. Once leaders have the necessary digital literacy, the transformative forward-looking vision, and the change-oriented mindset, they must also be able to deliver and decisively lead the organization into the future. In many ways, the biggest challenge facing digital leaders is simply having the will to make the types of changes necessary to adapt the organization to the emerging environment.

A good example here is Facebook's shift to mobile. As late as 2012, pundits questioned whether Facebook would be able to adapt its platform to a mobile environment. This question, of course, has now been resolved, since roughly 85 percent of Facebook's advertising revenue comes from mobile. Yet, this outcome was far from preordained. As Facebook sought to adapt its computer-based interface for the mobile environment in the early 2010s they chose to use HTML5 as the programming language. When it became apparent that the trend instead was to develop native apps, the engineering department was expected to retool and adapt to this new environment. The company was open minded enough to recognize when the technological environment was changing in unexpected directions and adaptable enough to make the necessary pivot in response.

What Traits Do Leaders Need More Of?

We asked what capabilities their leaders needed more of to help their organization navigate digital trends (figure 7.3). Somewhat surprisingly, these traits were similar across maturity levels.[3] We list only the top four here.

- **Providing vision and purpose.** Providing vision and purpose was the most desired trait of digital leaders. An aspiring vision and purpose serves as a compass to guide employees as they work, especially in distributed environments, where employees have greater autonomy to make decisions. Yet, it may not be sufficient to have that vision. Leaders must also provide the opportunity to execute that vision. George Westerman, of MIT, says, "To drive digital transformation, you need a very strong vision for where you're going and how it's going to be different. You then want to engage your people very strongly in owning and then fleshing out that vision. But then, third, you need to have very strong governance. What are the capabilities we're going to develop so that we can move transformations forward over and over again?"
- **Creating conditions to experiment.** This was the second most common trait respondents listed wanting more of. Consultant Ed Marsh notes how these conditions are being developed at a large food processing company. "First, we need to be hiring and selecting for people

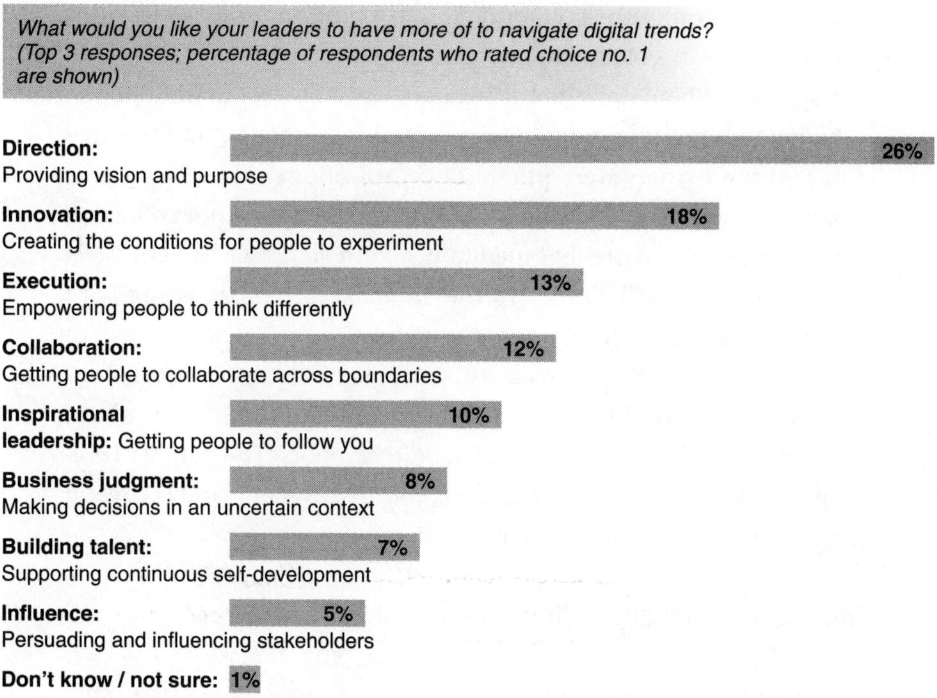

What would you like your leaders to have more of to navigate digital trends?
(Top 3 responses; percentage of respondents who rated choice no. 1
are shown)

Direction: 26%
Providing vision and purpose

Innovation: 18%
Creating the conditions for people to experiment

Execution: 13%
Empowering people to think differently

Collaboration: 12%
Getting people to collaborate across boundaries

Inspirational 10%
leadership: Getting people to follow you

Business judgment: 8%
Making decisions in an uncertain context

Building talent: 7%
Supporting continuous self-development

Influence: 5%
Persuading and influencing stakeholders

Don't know / not sure: 1%

Figure 7.3

who are more risk tolerant. Second, how do we create an organizational context where trying intelligently and not succeeding is okay, and may even be a behavior that we reward? Lastly, we want to put in place some platforms—virtual or physical—where people can play and experiment with new ideas and business models, including with other parties such as universities, entrepreneurs, etc."

- **Empowering people to think differently.** The third most desired trait was the ability to execute by empowering people to think differently. Thinking differently involves not only grasping what employees see as possible, but also understanding what customers expect and being prepared to respond accordingly. James Macaulay, of Cisco, says the company is sensing "that customers want to consume differently. That is, they want more cost value, experience value, and platform value. . . . We know we must change our stripes."

- **Getting people to collaborate across boundaries.** Last, getting people to collaborate across boundaries was also a common response. When we asked a separate question about the biggest barriers to collaboration in organizations, respondents indicated that those barriers were primarily organizational, such as culture, mindset, and silos. Our interview subjects thought about collaboration across boundaries in broad terms. John Halamka, of Beth Israel Deaconess Medical Center (BIDMC), spoke of collaborating with innovative organizations; Brent Stutz, of Cardinal Health, spoke of collaborating with partners; while Melissa Valentine, of Stanford University, spoke of the need to learn to collaborate with robots, which she calls "co-bots," or collaborative robots. The nature of collaboration that is necessary and possible in a digital world is beyond simple intraorganizational communication.

Creating a Culture of Distributed Leadership

CONF POINT!

One other key difference to leadership in a digital age: *where* leadership is found within the organization. In the twentieth-century hierarchical corporation, people looked only to the top of the org chart for leadership. With the pace of change, that is no longer practical—nor is it always where you can find effective leadership. When we talk about leadership, we are referring to leadership at *all* levels of the organization. Yet, we also observe a disconnect between the extent to which companies say they are pushing leadership down and how much they are actually doing it. While 59 percent of CEOs believe they are pushing decision making down, only around 33 percent of vice president and director-level respondents reported that it is happening.

Whether executives think they are pushing decision rights down but aren't executing, or lower-level employees are not stepping up to take those decision-making responsibilities is unclear. Chip Joyce, of Allied Talent, provides one perspective on this disconnect. He notes, "The middle organization—the first or second level of managers—often feel very, very differently about the organization than what the senior levels do. Trust isn't there. Belief that they have opportunities in the organization is lacking. Clarity of business strategy and whether the company has what it takes to succeed, the lower down in the organization you go[,] the shakier it becomes."

One characteristic of digitally immature companies is that leadership is trapped at the upper levels of the organization. This often leads to perceptions throughout the company that digital strategy is only talk, because—in practice—that strategy, no matter how carefully, articulately, or expertly it is formulated, never makes it out of the c-suite. To enable change, organizations must harness effective leadership at all levels of the company. As we discuss in subsequent chapters, digitally maturing organizations are less hierarchical and drive more decision making down to lower levels, where those decisions can be made more quickly and in a more informed way.

The Will to Transform

The will to transform is the last piece of the digital leadership puzzle. Organizations must make a conscious effort to transform and prioritize this transformation effort. Our research shows that whether digital transformation is a top management priority is a huge predictor of an organization's digital maturity—with 86 percent of respondents from maturing companies indicating that it is, compared with 34 percent of respondents at early stage companies. Respondents also said that a major barrier to digital maturity is simply that the organization has too many competing priorities. Strong digital leaders identify the changes necessary to adapt, and they focus relentlessly on them.

The time to start is now. The gap between digital capabilities and how companies operate grows wider each day. If your organization waits until evidence from the marketplace suggests that traditional business models are failing, it may be too late. Strong digital leaders see the disruptive trends in advance and make a conscious effort to do something about them. Successful digital transformation is an ongoing iterative effort, not a once-and-done project. It requires a flexible mindset and an organizational structure that allows the company to respond. Strong digital leadership is about helping keep the organization focused on the vision over the long haul and through multiple iterations.

Takeaways for Chapter 7

What We Know	What You Can Do about It
• Good digital leadership is needed at all stages of digital maturity. Leadership is found, and should be cultivated, at every level of the organization. • Providing vision and purpose and creating the conditions for people to experiment are key areas where managers can improve, in addition to empowering people to think differently and getting people to collaborate across boundaries. • Digital leadership moves beyond the command-and-control hierarchy of traditional organizations, toward cultivating vibrant networks that can act with higher levels of autonomy.	• Repeat the process described in the Takeaways for Chapter 6 across each level of the organization. • Conduct interviews and focus groups with lower-level managers, with the goal of understanding what they think will help them lead in more digitally mature ways. • Determine what actions can be taken on these findings (e.g., organizational and process changes, new technologies, training opportunities). • Ensure that the goal-setting and evaluation processes reinforce the cultivation of leadership traits that are appropriate for digitally maturing organizations.

*Not About Tech.
Talent & Mindsets*

"Our people are our most important asset." Eye roll. It's a well-intentioned sentiment that corporate leaders frequently utter, but one that has become so commonplace and banal that it's almost devoid of meaning or significance.[1] Nevertheless, great digital strategies first require great talent. In a world where change happens at an accelerating pace, an organization's best strategy is to establish the infrastructure that enables it to navigate the seas of change. And that begins with its most powerful assets: its people.

Technologies come and go. With decreased barriers to entry, technologies no longer offer a sustainable competitive edge. Organizations need to move with agility and think differently. They need to be creative and innovate. They need to anticipate, respond to, and create new value models. They can only do that with the right talent in place, and they need to think and act strategically about how to attract and retain that talent. They also need to understand what infrastructure, processes, and mechanisms to put in place that enable people to collaborate at scale, since much of the value that talent brings to an organization is unlocked when individuals interact with one another.

Our research suggests that the inability to attract and retain digital talent is, in fact, one of the most significant threats brought about by digital disruption. Before we can talk about how to attract and retain valuable talent, however, we must clarify what sorts of talent and skills are most critical for thriving in an increasingly digital work environment.

Hard Skills, Soft Skills, and Hybrid Skills

In a world of rapid technological change, technical skills and capabilities are critical and have become requirements for future jobs and careers. This is not a new challenge. Fifty years ago, the Soviet Union shocked Americans with its launch of the first orbiting space vehicle, Sputnik, months ahead of the then-humiliated United States. The event "sparked a much-needed revolution in scientific education in the US. America's scientific community, which had long been pushing for a new direction in science education, seized on the national mood to rejuvenate the curriculum."[2]

Today, we are experiencing our own version of the Sputnik effect. Indeed, the increased investment in STEM (science, technology, engineering, and mathematics) education over the past two decades speaks to a recognition of the important role these fields play in individual, organizational, and regional competitiveness. Programs, initiatives, and organizations like Girls Who Code crop up with regularity to encourage children to pursue these disciplines; and coding camps, robotics camps, and even K–12 STEM schools address the demand parents have for their children to be prepared for the future. Strong data support this investment in STEM, as technology is among the fastest-growing sectors for jobs in this country. The Bureau of Labor Statistics projects employment in these areas to grow to 9 million between 2012 and 2022.[3] Our research indicates that organizations cite technical skills, such as programing, data science, and analytics, as the skills in highest demand. Yet, while hardcore and deep technical skills are important and necessary capabilities for the future, they are not the only skills needed—and may not even be the most important skills needed.

Although some people include the soft sciences, like anthropology, psychology, and sociology, in STEM disciplines, emphasis has been predominantly on the hard sciences, engineering, programming, and math. Even so, the important role that creativity plays in innovation is being increasingly recognized, as reflected in the STEAM movement, which advocates adding art to the traditional STEM disciplines.

Companies are now starting to look for a balance of hard and soft skills, technical and business skills—all in the same person. Think of this as a "stack" of skills, analogous to the stacks that many organizations try to create with various technologies. We are seeing a rise in these hybrid roles in organizations.

Dan Restuccia, of Burning Glass, a Boston-based analytics company focused on identifying skill demands of the marketplace, elaborates on this phenomenon: "Companies are looking for people with a balance of technical and soft skills. Often people with deep knowledge in a certain area (e.g., technical, business) are also expected to have foundational skills such as communication, storytelling, etc. For example, data scientists need to understand the business and be able to tell the story to interpret the insights. Certain skills/roles are going away; and instead, everyone is expected to have these skills (e.g., social media)." Restuccia observes that new jobs are emerging that require both technology and liberal arts backgrounds, such as user-experience designer. "This job is one part designer and one part psychologist," he says. "When companies are looking for user-experience designers, many will seek candidates with degrees in psychology or anthropology."

Going beyond Skills to Mindset

Companies need people who can change, grow, and be agile to help their organizations do the same. In our research, we asked respondents to describe, in their own words, the most important skills that employees should have to succeed in a digital workplace. The 3,300 open-text responses were sorted into categories and subcategories (figure 8.1). Forty percent of respondents reported that being "change oriented" is the most important skill for employees to possess. Roughly half of the responses in this category explicitly mention being "change oriented," with the phrase "open to" followed by various ideas often included in the description. Other associated characteristics include adaptable, flexible, agile, and innovative. Being change oriented was particularly important for respondents from large companies. As we discuss in the

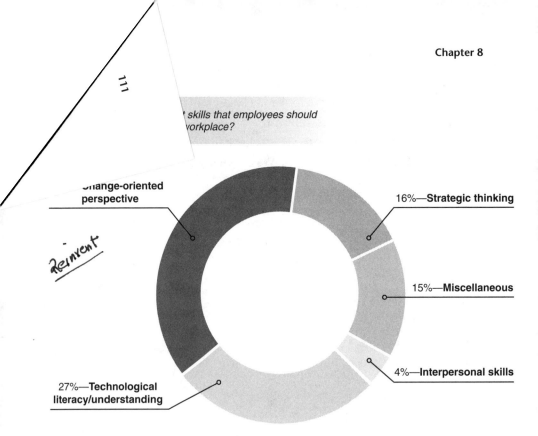

111

Reinvent

skills that employees should vorkplace?

Change-oriented perspective

16%—**Strategic thinking**

15%—**Miscellaneous**

4%—**Interpersonal skills**

27%—**Technological literacy/understanding**

Figure 8.1

previous two chapters, being change oriented is important at all levels of an organization, particularly among leaders.

One reason that being change oriented may be so important is that we are seeing an unprecedented rapid skill deterioration. A digital leader at a Fortune 500 company indicated that "In many areas of our industry, the expertise half-life runs about ten to twelve years. So, that is[,] if you learn something—if you were in sales, and then you left sales, you could come back ten or twelve years later and half of what you knew you could still use day to day in that job. In the digital area, I would pen the half-life at closer to eighteen months because our space is changing so fast." Employees recognize this shortening half-life for job skills. We asked respondents how often they needed to update their skills to do their job effectively in a digital environment (figure 8.2). Just under 45 percent said they needed to "continually" update their skills, with relatively little difference between respondents from companies across the

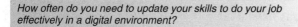

How often do you need to update your skills to do your job effectively in a digital environment?

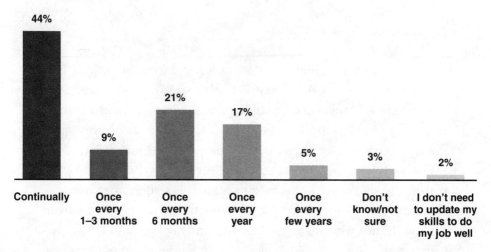

Percentages do not total 100 due to rounding.

Figure 8.2

maturity spectrum. Just over 45 percent said they needed to update their skills every year or more often. In other words, 90 percent of our respondents felt they needed to update their skills at least yearly. Employees understand that keeping their skills up to date is a pressing concern.

Chris Cotteleer, of Schumacher Clinical Partners, put the challenge of digital talent in perspective. Cotteleer's statement is indicative of a talent challenge that many organizations and leaders face: it's not enough to have specific skills or experience. "Do we have enough talent? No. Are we looking for more? Yes. Is it hard to find? Really good talent is. What do we look for? Vision is important and necessary, but not sufficient. You also have to have strategy and execution, and that means you have to have some sort of *je ne sais quoi*. There's a passion inside that I look for. Technology we can teach in a lot of the cases, or we can rent. I'm looking for people that can think." But having the right skills is not enough: today's talent must embrace and navigate change. To meet this shift in work, confront new challenges, and tackle emerging opportunities, organizations need a talent base with the right disposition and mindset.

A Continuous Learning Mindset

When we asked a partner at a leading Bay Area venture capital firm what one book someone should read to get a sense of how to succeed in Silicon Valley, he suggested *Mindset* by Carol S. Dweck. Dweck contrasts two types of mindsets—growth and fixed.[4] Her research reveals that mindset plays a much bigger role in success than innate talent does. People with fixed mindsets believe that intelligence (along with talent, personality, and other traits and capabilities) is static: either you have it or you don't, and you can do little to change it. Having a growth mindset begins with the core belief that intelligence (and talent, personality, capabilities) *can be developed*—it is not static or predetermined. For people with growth mindsets, the focus is on the *process* as well as the outcome. The importance of process and effort are critical, as they affect *how* and *what* is learned and the progress made. In much the same way that digital leadership is a learned skill, so is the ability of employees to work effectively in a digital organization.

For digital talent, this difference in mindset is often a key component of success, and we believe that a fixed mindset can describe both an organization as a whole and the individuals in that organization. Evidence of a fixed mindset abounds in many companies. One trait of people with a fixed mindset is that they tend to assign labels to themselves (and to others) based on IQ, test scores, and early performances and results: smart, dumb, talented, winner, loser. Likewise, those with a fixed mindset who lack digital skills often refer to themselves with similar sorts of labels— "I'm just not a digital person," or "I'm technologically challenged."

A key facet of developing digital talent is cultivating a growth mindset. Of course, no matter how much of a growth mindset you nurture, not everyone in your organization is capable of learning advanced technical skills, like Hadoop or machine learning. Yet, everyone can become more digitally literate, learn to adapt better to change, and think more critically—the skills our survey respondents identify as most important for success. The growth mindset is critical to employees' ability to continue developing the skills and knowledge necessary to work effectively in a rapidly changing environment. This mindset involves individuals who typically embrace challenges, persist in the face of setbacks, see effort as a path of mastery, learn from criticism and feedback, and find lessons and inspiration in the success of others.

Donna Morris, of Adobe, notes that these are the type of people she seeks to hire: "If I step back, our industry is one that's all about change. So that hasn't changed. We've always based hiring on looking at individuals that were continual learners. We call that learning agility—they demonstrated a lot of learning agility; they were intellectually curious." Perry Hewitt, former chief digital officer at Harvard University, says, "I always say hire agility over skills. And interest and aptitude over demonstrated track record."[5] A person's skills, experience, or even track record can no longer be indicators of future success.

"Talent" is defined as an innate or natural aptitude or skill. The irony is that digital talent is less about individuals' innate skills and aptitudes, or what they can do today, than about *what they will be able to do tomorrow*, based on their mindset and their ability to learn and grow. But this type of learning differs from institutional learning, which often takes place in classrooms and training programs. It is necessarily more self-driven and less structured. Institutional learning and formalized training alone are unable to keep up with the pace of change and technological development. A different type of learning is needed: one that is continual, experiential, and exploratory. It is about finding and seeking learning in nearly everything—being in a constant state of growth.

Technology Is Not Just about Millennials!

Nowhere is the lack of a growth mindset more apparent than in the preoccupation of business leaders wanting to connect digital maturity with millennials. The belief in popular culture is that somehow millennials

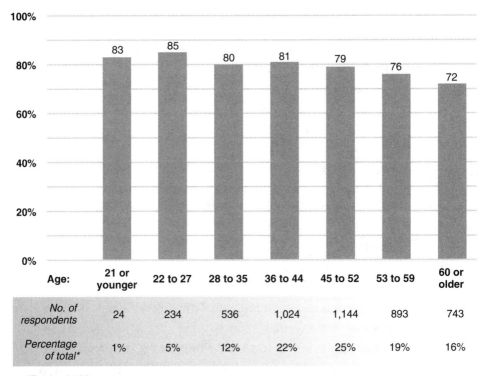

How important is it to work for an organization that is digitally enabled or is a digital leader? (Respondents who answered "very important" or "important")

Age:	21 or younger	22 to 27	28 to 35	36 to 44	45 to 52	53 to 59	60 or older
No. of respondents	24	234	536	1,024	1,144	893	743
Percentage of total*	1%	5%	12%	22%	25%	19%	16%

*Total = 4,598

Figure 8.3

are born or naturally imbued with the need and/or ability to be digital. Across the four years of our research, however, age is a surprisingly weak predictor of digital perceptions and desires. Consider figure 8.3, which shows a roughly 10 percent difference across age groups reporting how important it is to work for a digital leader.

Certainly, there is some difference between the youngest and oldest respondents in our sample, but not the radical departure you might expect. Even a surprising 72 percent of people aged sixty or older said it was important for them to work for a digital leader. We find this trend to be generally consistent across all questions we asked. The variations across respondents may not necessarily relate to when they were born or the environment in which they grew up. Instead, the differences might reflect the social environment they find themselves in now, differences that might change as they mature. Admittedly, our survey on digital disruption might appeal to a more digitally minded group, which could bias our data somewhat on this point. Nevertheless, in our experience teaching undergraduate, graduate, and executive students over the course of a decade, age is far less of a predictor of digital literacy and aptitude than commonly believed.

In fact, we have often found it easier to teach older workers the digital literacy they need to thrive than to teach younger workers the organizational knowledge they need. Younger students tend to excel in the *procedural* applications of technology; that is, they are good at navigating the various apps and platforms. Older students, however, tend to be superior in the *strategic* applications of technology. Once they are familiar with the capabilities of technology, they tend to appreciate the business applications more quickly than do younger students.

The key takeaway here is that millennials are not inherently digital, at least not in an organizational sense. They may have adopted technology individually, but they will not instinctually know how to help your company adapt. Even if they come out of college more digitally minded than their older counterparts, that edge will atrophy quickly without continuous learning and a growth mindset, because the technology keeps changing. Neither are older workers at an unresolvable disadvantage. Older workers may not have the time or inclination to become hardcore data scientists, but they can acquire the necessary digital literacy to contribute productively to their digitally maturing organizations. Millennials certainly have much to teach about technology, but they also have much to learn about organizations and business. Developing a growth mindset in your organization frees up employees to develop long-term digital talent, regardless of their age.

Cultivate an Organizational Growth Mindset
Aligned with Your Digital Strategy

The organization itself can develop a growth mindset, not just the people in it. If your company is typified by a fixed mindset, you may need to change these labels first to get people to think differently and to cultivate the skills they need to succeed in a digital organization. An interesting trend in our research is that several executives said they need to "trick" their employees into digital initiatives. If they talk about their transformation efforts in terms of digital, the fixed mindset takes over—employees don't believe that they or their organizations can do it because they just aren't digital people. Instead, they talk about efforts in terms of improved customer service or exploring new ways to connect, without explicitly referencing the digital aspect. The organizations and the individuals in them are trapped in a fixed mindset.

Organizations must create the kind of environment where employees want to stay—but also where ideas can flourish. Given the changing skill-set landscape and rapid skill deterioration, individuals and organizations both need to build a culture of continuous learning and foster and encourage growth mindsets, which are key to continual evolution and adaptation. Organizations with growth mindsets emphasize learning as an organization and as individuals. They encourage experimentation and pilot programs, and they foster creativity and innovation. Studies have linked growth-mindset organizations with increased innovation, collaboration, and risk taking.[6]

Characteristics of growth-mindset companies align with attributes of maturing digital organizations. Supervisors tend to rate their employees more positively, saying that they are "more innovative, collaborative, and committed to learning and growing. They are more likely to say that their employees have management potential. . . . At a minimum, growth-mindset firms have happier employees and a more innovative, risk-taking culture. . . . Growth-mindset organizations are likely to hire from within their ranks, while fixed-mindset organizations reflexively look for outsiders."[7] (Fixed-mindset companies also focus particularly on pedigree as a desirable employee characteristic.) An example of growth-mindset hiring practices is how Google "has recently begun hiring more people who lack college degrees but have proved that they are capable independent learners."[8] The good news for organizations is that mindset can be learned, as research has demonstrated.

Developing Digital Talent at Beth Israel Deaconess Medical Center

When working in IT, keeping abreast of the latest and changing technologies can be challenging. BIDMC's CIO, John Halamka, has managed to build and develop an edgy and innovative team—despite being underfunded. Halamka identifies various strategies for developing skills and capabilities in his people:

1. **Develop existing personnel:** Talent development consists of internal and external training courses, boot camps, and conferences, as well as providing new opportunities for employees to grow. For example, Halamka had a web team focused on traditional web development. But since 80 percent of the access to BIDMC's software applications were through mobile devices, they had to change. "After a search, we found that Kony is this application development platform that enables Android and iOS apps to be developed once and repurposed across multiple OSs. And so what we did is we licensed that stuff. But then we also licensed a boatload of services, and we ran internal boot camps and internal hackathons and those kinds of things, and we converted our entire web team into a mobile team without hiring a single new person. . . . By and large, a few of the teams say this is a fabulous opportunity, let's jump. Then the others say, oh, I'm not quite sure, but I'll go along for the ride and eventually come around; and those who never come around, and those [who] depart."

2. **Establish co-op programs with local colleges:** "We will bring in a bright young person and[,] through an internship experience, develop them to the point where we very often hire them when they graduate. And so that's a sort of variation on the internal model."

3. **Learn from collaborative partnerships:** "I created something called the Center for IT Exploration, or Explore IT, for short. And basically, it is very analogous to the 20 percent thing that Google used to have, right? Take your best people through a meritocracy and ask them to spend 20 percent of their time investigating something outside the scope of their day job. Oh, I've just discovered this cool thing that Google does. Oh, great. Let's see if Google would collaborate with us. . . . Lots of interesting lessons learned. And that's sort of the way it works. So, we have a center that

(continued)

> basically is permissive of these kinds of collaborations. And then once you establish the Google and the Amazon and the Facebook or the other third parties, then you can get really remarkable ongoing synergy."
>
> 4. **Recruit externally for targeted skills:** "Only very rarely do we do what I'll call the Monster.com search for somebody who already has the skills. Sometimes. I mean, it's rarified, but if there's a particular domain where it's just extraordinarily challenging to teach someone Oracle from the ground up, you know—you'll pick somebody from another organization and promote them, or that kind of thing."

Don't Just Train; Create Opportunities to Learn

An organizational growth mindset also suggests that organizations should move beyond just training to develop the skills employees need for the future. We have found that the skills required to thrive can be learned by and taught to people of all ages. What's more, research shows that intrinsic motivators are quite powerful, and people have a need for autonomy, growth, and meaning in their work.[9] In other words, people want to learn and grow. Our own research shows that digitally maturing companies address this need by providing the resources or opportunities to help their professionals develop skills to thrive. Through investment in talent and carefully cultivated cultures, digitally maturing companies are able to develop the skills and capabilities they need while simultaneously creating the conditions for meeting employee needs and intrinsic motivators.

If the desired skills are a blend of technical and soft skills, formal training classes in topics such as coding or data analytics are unlikely to be the solution—although they may be part of one. Rather, digitally maturing organizations provide diverse environments in which these employees can develop the types of characteristics and skills they reported as being important—openness, adaptability, flexibility, agility, and innovation. Professor Prasanna Tambe, of the Wharton School, emphasizes this distinction when he notes that learning has to "go beyond training. Given the pace at which technology is moving, it's difficult to create and reproduce and administer training in these different technologies. A lot of it's done on the job, and the learning environment is important. Firms are focusing on creating the types of environment [where] workers can teach themselves and creating physical environments where workers like to spend time."

Tambe notes that some companies are increasingly allowing their employees to dedicate work time to open source software communities. Contributing to open source code provides several benefits to workers. They can continue to gain skills and credibility within the open source software community, which can then be brought back to the organization to inform the company's ongoing efforts. It also enables continual skill development, which means employees are less likely to leave the company to learn or improve skills. Many companies find that paying their employees to work for these open source communities delivers a positive return in terms of digital maturity.

John Hagel, from Deloitte's Center for the Edge, suggests that technology can also be an enabler for this skill development: "Our view is that the learning that's going to be most powerful is not by accessing what other people already know. It's driving new knowledge creation through practice in the workplace, rather than in a training room, by addressing challenges and business situations that have never been confronted before. That's a very different approach to learning." A good example of an organization that creates this type of environment is Salesforce. The company has created a platform called Trailhead, in which employees can both teach and learn from one another. Employees can participate in or create learning modules or series of modules demonstrating particular skill development. The platform records what learning opportunities they have taken advantage of, which is factored into their performance evaluations. The platform also indicates which employees have created modules that are most used, which is typically a point of pride for these creators.

Perhaps leaders in digitally maturing organizations should consider two additions to the standard list of questions they pose to job candidates.

ıve you learned lately that makes you a more competent ınployee in a digitally maturing organization? Second, how ıed someone else to learn something that helps them to ...... a more competent and capable employee in a digital maturing organization? In today's business environment, we can ill afford to hire individuals who can't offer up compelling answers to both questions.

Takeaways for Chapter 8

What We Know	What You Can Do about It
• The most important skills for succeeding in a digital environment are strategic thinking, change orientation, and growth mindset.	• Embed learning plans into your talent management processes, but ensure that learning extends beyond formal training programs (e.g., job rotation).
• Most respondents are dissatisfied with the opportunities their organization provides to develop relevant digital skills.	• Identify opportunities for employees to help others learn, and reward employees who are effective learning enablers.
• The trend of continual learning is not driven just by millennials. Employees of all ages reported a desire to work for digitally maturing companies that allow them to continue to develop their digital skillset.	• Include the capacity to learn and to help others learn as an explicit part of your recruiting and evaluation criteria.

LEAVE Behind Questions

9   Making Your Organization a Talent Magnet

In 1971, Alice Waters and a small band of friends founded a neighborhood bistro called Chez Panisse. It was named for Honoré Panisse, the most generous and life-loving character in Marcel Pagnol's 1930's film trilogy about waterfront life in Marseille . . . as an homage to the sentiment, comedy, and informality of these classic movies.
—"About Chez Panisse"

Throughout its existence, Chez Panisse has focused on ingredients, rather than technique, cultivating a supply network of direct relationships with local farmers, ranchers, and dairies. That the restaurant has survived for so long is remarkable. That it has won multiple awards, ranging from Michelin stars to being named by *Gourmet* magazine as the best restaurant in America, is impressive. That it has been at the forefront of culinary innovation is even more of an achievement. But what is perhaps most noteworthy is its ability to attract some of the top culinary talent in the United States.

Chez Panisse's alumni list reads like a who's who of famous chefs and restaurant owners, ranging from April Bloomfield, of Spotted Pig fame, to Jeremiah Tower, of Stars, the "inventor" of California cuisine.[1] Dartmouth professor Sydney Finkelstein tells the story of Alice Waters, fashion iconoclast Ralph Lauren, Oracle founder Larry Ellison, producer George Lucas, SNL creator Lorne Michaels, legendary football coach Bill Walsh, and hedge fund manager Julian Robertson in his book

Superbosses: How Exceptional Leaders Master the Flow of Talent. What they all have in common is that they are "legendary for spawning legions of protégés who have gone on to transform entire industries."[2] Organizations that aspire to digital maturity need to be talent magnets, just like the organizations, headed by Waters and other icons, that Finkelstein describes. What do these icons do? They have high standards, a prodigious appetite for coaching, recognition that they also benefit from the coaching of others, a willingness to take smart risks, and the ability to deconstruct complicated activities into component parts that can be learned and mastered.

In the previous chapter, we show that digitally maturing companies do a much better job than less mature companies of developing the digital skills of their employees. Unfortunately, developing existing skills alone is necessary but not sufficient for competing in the future. You must not only train your existing talent, but also attract and retain new talent. Organizations must address these challenges without falling prey to well-intentioned, but largely symbolic, actions that fail to address access to this scarce and highly mobile segment of the workforce.

Given that a lack of digital talent appears to be a problem for most companies seeking to mature digitally, we would expect to see companies identify access to talent as a major risk. Yet, when we questioned respondents about the biggest risks they faced in responding to digital disruption, access to talent barely registered. Finding the right talent is one challenge; keeping that talent can be equally challenging, particularly if an organization is not providing the right signals to employees about the significant role digital business plays in its overall company strategy. In this chapter, consequently, we share our perspectives on what it takes to become a talent magnet.

First, Make Good Use of the Talent You Have

Our research suggests that digitally maturing companies do a better job than less mature firms of developing the talent they have. Figure 9.1 illustrates the strong relationships among responses to three different

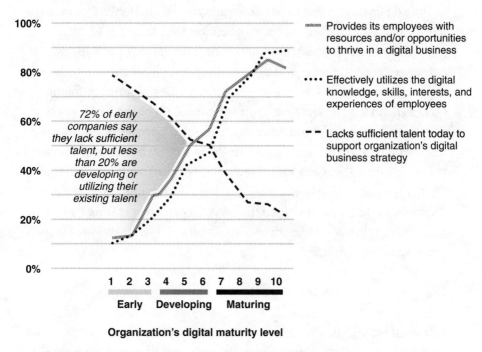

Commitment to digital talent and digital maturity. My organization:

- ‖‖‖‖ Provides its employees with resources and/or opportunities to thrive in a digital business
- •••• Effectively utilizes the digital knowledge, skills, interests, and experiences of employees
- – – Lacks sufficient talent today to support organization's digital business strategy

72% of early companies say they lack sufficient talent, but less than 20% are developing or utilizing their existing talent

1 2 3   4 5 6   7 8 9 10

Early   Developing   Maturing

Organization's digital maturity level

Figure 9.1

questions in our research: (1) Does your company provide its employees with resources and/or opportunities to thrive in a digital business? (2) does your company effectively utilize the digital knowledge, interests, skills, and experience of employees? and (3) does your company lack sufficient talent to support your organization's digital strategy?

The relationships among these questions played out about as you might expect, but the differences were extremely stark. Between 80 percent and 90 percent of respondents from digitally maturing companies said their businesses provide resources to thrive and effectively utilize the skills of employees, but only 20–30 percent of respondents from early stage companies said the same. Likewise, 70–80 percent of early stage companies reportedly lack sufficient talent to support their digital strategy, while only 20–30 percent of maturing stage companies do. As with leadership, the difference between early and maturing companies is not whether they have enough talent, but what they are doing to develop it. Since all levels of talent need to increasingly carry leadership responsibilities, developing talent is even more critical for companies to do.

Digitally maturing companies spend time developing the skills their employees need to thrive, and in turn, these employees effectively help those companies execute their digital strategy. Less mature companies do not spend time developing or utilizing the skills of their employees, and—surprise!—these companies don't seem to have sufficient talent to execute their digital strategies. These data reflect the old story about two executives discussing employee training. The first asks, "What if we train our employees and they just end up leaving?" The second responds, "What if we don't and they stay?" Digitally maturing companies invest in the development of their employees' skills, and they reap the rewards for doing so.

Don't Lose the Talent Once You Have It

While attracting, retaining, and developing talent are distinct challenges, these challenges are related. Respondents from maturing companies that offer employee training still reported having only *sufficient* talent to execute their strategies; training does not prevent them from wanting to attract still *more* talent. That companies need to attract even more talent to compete effectively in a digital world is not surprising. What did surprise us was finding that this need appears to transcend digital maturity level. While over 70 percent of respondents in the least mature companies reported the need for new talent, more than 50 percent of digitally maturing companies indicated a similar need. All companies want and need more and better digital talent, regardless of how well they train the employees they already have.

Talent leakage compounds the challenge of recruiting people in the first place. As we discuss in chapter 8, people clearly prefer to work for a digital leader. Many employees seeking to leave a company are frustrated

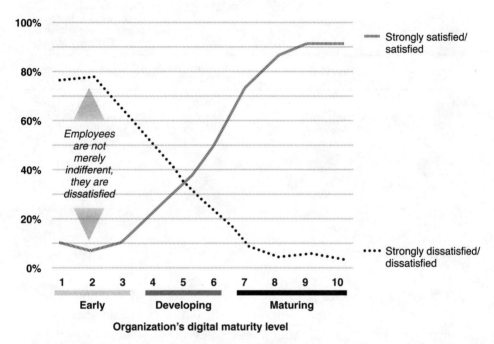

I am satisfied with my organization's current reaction to digital trends (Respondents who answered "strongly agree" or "agree")

Figure 9.2

by their organization's lack of response to digital trends. While 81 percent of surveyed employees from digitally maturing companies are satisfied with their organizations' response to digital trends, only a paltry 10 percent of employees from early stage companies reported satisfaction (figure 9.2). Even among employees in developing-stage companies, only 38 percent said they are satisfied. So, not only do employees want to work for digital leaders, they may start looking to leave if their companies do not actively seek to mature digitally.

In our research, the number one way to reverse the talent drain was to provide employees the opportunity to grow and develop. For respondents from companies that provide opportunities to develop skills for working in a digital environment, the desire to leave drops precipitously.

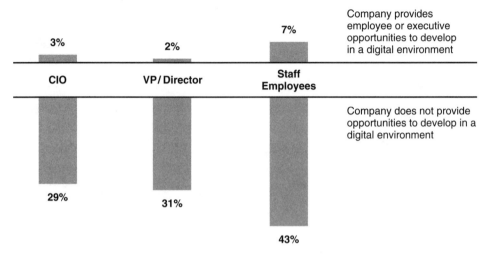

Percentage of respondents who plan to leave their companies <u>in less than one year</u> given digital trends (Specific roles shown for illustration)

7% — Company provides employee or executive opportunities to develop in a digital environment

3%   2%

CIO   **VP/Director**   **Staff Employees**

Company does not provide opportunities to develop in a digital environment

29%   31%

43%

Figure 9.3

Although similar results are evident across different levels of employees, we were surprised to find them particularly pronounced among middle management. VP and director-level employees were *fifteen times* less likely to report wanting to leave their companies within a year as a result of digital trends if their employers provided opportunities to develop skills necessary for working in a digital environment (figure 9.3). CIO positions and IT and sales staffs were also significantly more likely to report wanting to leave within a year if their companies lacked opportunities to develop these skills. Furthermore, providing opportunities to grow and develop increases the likelihood of staying ahead of the skills deterioration, simultaneously cultivating a growth mindset.

Yet most companies aren't supporting their employees' desire to continually update their skills to do their jobs effectively, and these results are relatively robust across age groups. Only one-third of respondents said they are satisfied with how their organizations are helping them prepare for the changes necessary to work in a digital environment (figure 9.4). As might be expected, these results vary by maturity. While

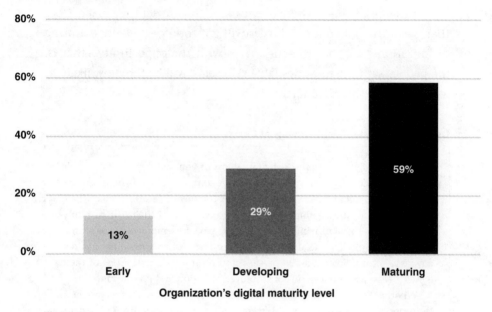

I am satified with how my organization is helping me prepare
for the changes necessary for working in a digital environment
(Respondents who answered "strongly agree" or "agree")

Early — 13%
Developing — 29%
Maturing — 59%

Organization's digital maturity level

Figure 9.4

59 percent of respondents from digitally maturing companies reported being satisfied with how their organizations are helping them prepare, only a paltry 13 percent from early stage companies are satisfied, and the vast majority of these respondents are actively dissatisfied.

In chapter 8, we talk about intrinsic motivators and people's need for growth, autonomy, and meaning in their work. Digitally maturing organizations are most likely to address those needs through their investment in developing people and their flattened hierarchies, which drive more decision making down into the organization (something we discuss more in part III). Investing in talent and giving people opportunities to develop address the need to continually grow and learn. Distributed leadership allows for more decision making and ownership without always having to "run it up the chain." And while meaningful work or purpose is different for everyone, it helps to have a clear and coherent digital strategy that is tied to the overall corporate strategy (another characteristic of digitally maturing organizations).

Put in other terms: People want to continue to learn and to be part of exciting new things, where they can see the impact they make. Does that mean that all your employees will go work for digitally maturing organizations? No, but the employees with the opportunity—that is, the people with the most desirable skill sets—will be the most likely to leave sooner rather than later.

Developing a Digital Talent Strategy at Cigna

If you are going to develop the digital talents of your employees, you should know what talents you want to cultivate. A company among the most strategic about its talent development model is the health insurer Cigna, which aggressively encourages employees to take advantage of their tuition remission program, through which the company pays for employees to continue their education. Cigna leaders analyzed the effect of this program and discovered that it delivered a 129 percent return on investment. Most of these benefits were realized in terms of employee retention and promotion.

What makes Cigna unique is the strategic way it approaches this program. Its leaders determined which skills were going to be most critical for the company in the coming years, then identified roughly a dozen talent areas deemed strategic skill sets for the company's future. Employees that pursue a degree or certification in one of these strategic areas receive tuition reimbursements at roughly three times the rate of other degree programs. Thus, Cigna is developing talent in strategic areas by providing the resources for their employees to obtain that training on their own. Through the program, the company can align education reimbursement with its talent pipeline needs as well as the needs of its employees to maintain employability.

Passive Recruiting Exacerbates the Talent Threat

The talent news gets even worse for less digitally mature companies. Not only are employees more likely to leave if they don't obtain the opportunities to continue developing, but digitally maturing companies are also more likely to come get them. Less digitally mature organizations face an emerging practice known as *passive recruiting*, when companies search LinkedIn or other professional platforms to identify and approach individuals who have skills they desire, even though these people are not actively seeking to move. An executive we talked to notes: "The digital leaders that we bring in are not leaders that typically are looking for us. We're typically looking for individuals that come with scale that have likely worked—and this is at the leadership level in particular—they've likely worked for companies that are larger than us, that are global. They are used to degrees of complexity and . . . can roll up their sleeves and sort of make things happen."

Digital platforms enable this passive recruiting trend. Indeed, 75 percent of respondents from companies at all levels of the maturity spectrum reported that digital platforms had raised their profiles outside the organization. Over 50 percent of respondents had been approached unsolicited via these platforms by companies offering intriguing job opportunities. You can assume that the employees approached in this way will represent the more valuable half of your organization, not the less valuable or less productive employees.

Figure 9.5 graphically represents the recruiting advantage that digitally maturing companies possess. We tested two statements with survey respondents: (1) My organization's embrace of digital business attracts new talent; and (2) my organization needs new talent to compete in the digital economy. Note that the intersection between the two lines in figure 9.5 is above the developing category. This result suggests that even developing companies are at risk of their best talent fleeing to digitally maturing companies. Forewarned is forearmed. Digitally maturing companies know that they possess an advantage in recruiting, and they are coming after your most valuable employees.

Maturing companies are not going to simply remain satisfied with their current talent base—indeed, over 50 percent of these companies' respondents said they still need new talent—and they will try to recruit the best employees from your company. They want more and better digital talent, and they know that employees want to work for companies like them. When we asked respondents how their organizations are developing digital talent (figure 9.6), the top answer for both

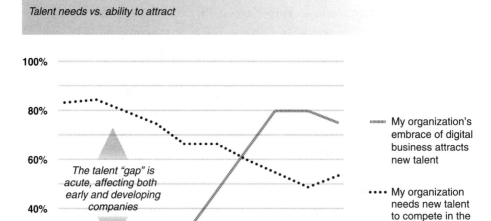

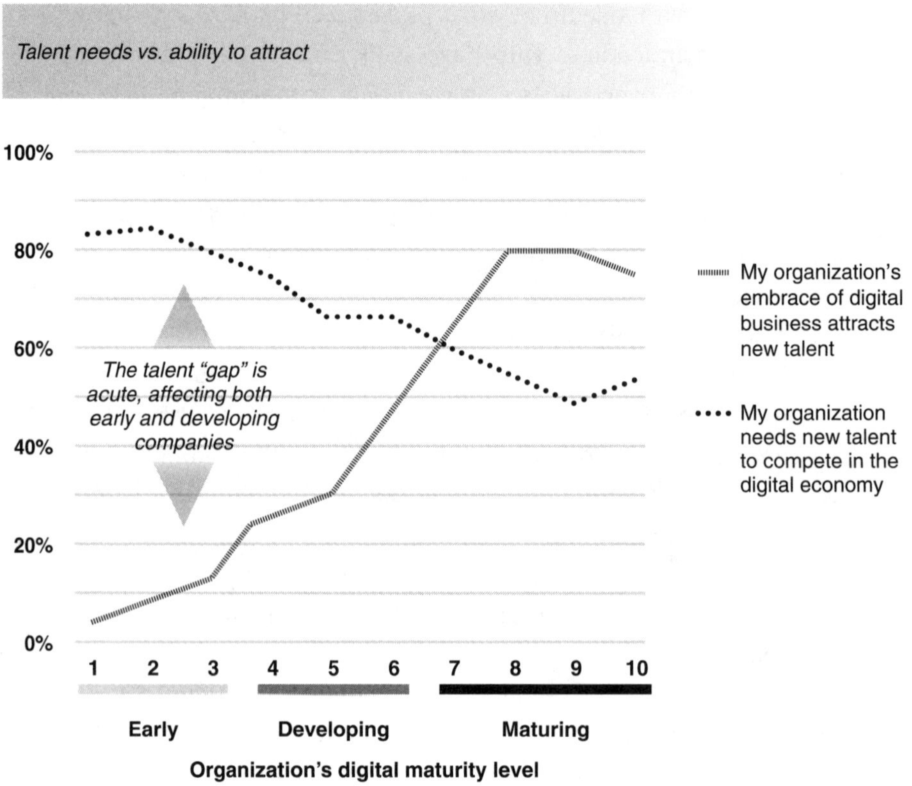

Figure 9.5

	Early	Developing	Maturing
How are companies primarily strengthening digital innovation capabilities?	**1.** Hiring contractors/ consultants **2.** Don't know **3.** External relationships **4.** Developing employees	**1.** Developing employees **2.** External relationships **3.** Hiring contractors/ consultants **4.** Recruiting digital employees	**1.** Developing employees **2.** Recruiting digital employees **3.** External relationships **4.** Recruiting digital leaders

Figure 9.6

maturing and developing organizations was that they cultivate digital talent within their existing employees. Yet, the second most important way maturing companies develop digital talent is by recruiting new employees. For developing companies, recruiting new hires is only the fourth most important method. Digitally maturing companies recognize the advantage they have in hiring talented employees, and they are prepared to leverage that advantage to attract more talent.

Locate Where the Right Digital Talent Can Find You

The three most important words in real estate are "location, location, location." Location may be equally important for developing digital talent. Digital talent may seem relatively mobile, given the availability of platforms through which work can be done. Yet location seems to be a big factor in talent markets, where employees with valuable skills move between companies. Silicon Valley is likely the most famous location for strong digital talent markets. Many companies have sought to establish innovation hubs there for just this reason.

Prasanna Tambe, of Wharton, suggests carefully considering the choice of location, because you need to balance attracting with retaining talent when considering Silicon Valley. He says,

> When you're located in an area like that where there are a lot of skilled workers[,] you also have this retention problem. It's difficult to keep workers, because they can always move on to other places. So, there's some degree of strategic decision making[,] where firms are choosing how to separate their workforce operations in a way that they can balance the benefits of working in a place like Silicon Valley. If they're hiring in a place like Silicon Valley, they may not get the retention benefits of working in labor markets that are not as tight and which may not have the same retention costs as other parts of the country.

Furthermore, the types of skills you are looking for factor into the decision. If you are looking for talent with cutting-edge capabilities, then moving to Silicon Valley may be the best option. If you are seeking to develop more general tech skills, then other places offer great value.

Tambe notes that university towns often provide the type of talent development companies are looking for—at a great value.

Look Outside Your Industry for Fresh Perspectives

Expand your search for talent beyond your industry. One way to do this is to hire leaders from other industries who can begin to infuse the company with a digital mindset. These leaders are often called "anchor hires," because they attract digitally minded employees who want to work for them. Common anchor hires are business leaders who possess both business and technical skills. One executive at a large consumer products company notes,

> Before, if you wanted to get to senior levels in an organization like ours, you [had] to work in other organizations that look like ours. And they all operated relatively the same. Well, now you have an entire workforce of digitally fluent talent that works on businesses as big if not bigger than the businesses that we run. They all run P&Ls, they all have had the same political experience. They've all figured out how to maneuver things through bigger organizations, be it Amazon, Google, Facebook, Cisco, Intel, or any of these. And they are beginning a migration over to organizations that look like us. And it's very difficult for current talent to be able to compete, because while they all have the same level of organizational intelligence, they lack digital fluency.

When adopting this approach, it is important to emphasize what's unique and worthy in your sector. Kristin Darby, of Cancer Treatment Centers of America says,

> In my opinion, health care is probably one of the most rewarding industries you could ever work in, because we're affecting people's lives every day. If that truly resonates with someone and they have the dynamic skill sets to be a visionary, understand how mobile and digital can change experiences, and they can resonate with patients, I think that that's really what we're looking for, which is very difficult to find. I also have tapped talent from other industries farther advanced than health care—financial services, for example, in the digital area. A recent head of our consumer technologies that joined us led mobile for a global financial services firm, for example. Bringing in that type of talent really allows us to leverage knowledge, and really the learnings that many other industries have experienced, in new ways.

Leverage Talent beyond Your Four Walls

Perhaps the most consistent strategy across all maturity levels is to work with external relationships to develop digital capabilities. This approach portends a shift in the marketplace to talent ecosystems. The disruption caused by technology is so pervasive and fast moving that companies sense they can no longer afford to go it alone. They cultivate a network of partners who learn together how to change in response to digital trends. For example, Cardinal Health has created a digital innovation center, where customers and business partners are routinely invited to collaborate and help drive innovation. Brent Stutz, director of Cardinal Health's Fuse center, found that only by understanding the needs and situations of these stakeholders could he really develop the type of innovation the company sought (see chapter 13).

These external partnerships may also consist of unconventional relationships, for example, with online communities, such as the coding community Topcoder. Companies are increasingly turning to platforms like these for access to on-demand digital talent. Mike Morris, CEO of Topcoder, explains,

> Firms have all come to the conclusion that the workforce really is changing. So, people have been talking about it for years, and the next generation workforce and what's going to happen. But when you're a company that's constrained by the number of quality people you can hire to your growth, and you start seeing that one out of three Americans are freelancers—that's what they're choosing as their career path. That's getting more and more popular. It's going to get harder and harder for them to compete with the talent, because people don't want to go and sit at a 40 hour a work week job where they're told what to do and they're put on assignments. They'd much prefer to have the freedom.

Platforms also create opportunities to retain on-demand nondigital talent as well. For example, the restaurant-ordering company Olo's Dispatch platform, described in chapter 4, taps into the Uber network to provide on-demand food delivery drivers.[3] This platform also uses analytics to help the restaurant know when to start cooking the order, based on driver availability, as well as the delivery radius to maintain food quality under current traffic conditions.

Encourage Talent to Leak within Your Organization

Earlier generations of management scholars advocated "managing by walking around."[4] This common-sense management style has endured through generations since. "Management by walking around (or MBWA) is the habit of stopping by to talk with people face to face, get a sense of how they think things are going, and listen to whatever may be on their minds."[5] The philosophy can be extended to digital skill development. Bee pollination is often used as a metaphor to describe how knowledge moves in many organizations. Just as bees move pollen by flying from flower to flower, employees carry knowledge as they move between different assignments. At each stop, employees pick up new skills and knowledge and then leave behind for others something they have learned. Companies are increasingly adopting a "tour of duty" model, in which employees spend a fixed period in a certain assignment, then move on to another assignment, which may be entirely different.

This model differs in two key ways from how talent is managed at most organizations. It does not assume that a person will stay in a job indefinitely, and it does not necessarily move employees vertically within the same function. The tour of duty model imposes on employees the need to continually develop diverse skills, and it allows for this diversity to come from a career within the organization rather than necessarily going to another company to advance. Furthermore, the process helps people learn to become more adaptable to change as they are repeatedly thrust into new environments.

This moving around might also involve people not explicitly taking new jobs but simply being exposed to new experiences through new projects, new teams, and new challenges. Greg Baxter, chief digital officer of MetLife, echoes this sentiment, noting "you have to act your way into new thinking, not think your way into new acting. Much of what we're doing is trying to create the programs and the opportunities for people to become comfortable, confident and competent in doing different things and doing things differently." He notes that MetLife has "a real commitment to the people that we have in the company to make sure that they have the skills and opportunities to thrive in a digitally disrupted world." The company has launched a $10 million investment in its global workforce—a program it's calling the Workforce of the Future Development Fund. The program will be available to all employees, and includes a special learning program on digital skills. The skills MetLife hopes to see employees develop are an ability to work with an increasingly diverse workforce, an ability to work comfortably with next-generation technologies, and an awareness of new business models that involve platforms and ecosystems.

We started this chapter with the story of Alice Waters and Chez Panisse. As Finkelstein notes in *Superbosses*, Waters's success in attracting some of the best culinary talent in the United States is not an accident but the result of individual and institutional practices and habits. While the world of digital disruption presents special challenges in attracting talent, many of the habits transcend digital, beginning with the need to provide employees with genuine opportunities to acquire the skills and experiences that will help them succeed and move into roles that take advantage of these newly acquired skills and experiences.

Takeaways for Chapter 9

What We Know	What You Can Do about It
• Developing your employees is a necessary, but not sufficient, means of securing talent required to compete in a digital world. Acquiring and retaining talent are also key.	• Develop a long-term strategic plan for the types of talent your organization will need to compete in a digital environment and take active steps to retain that talent.
• Developing and retaining talent are interrelated. Employees reported being up to fifteen times less likely to leave their companies within a year if they received developmental opportunities.	• Be sure your most valuable employees have the time and opportunity to take advantage of opportunities to develop digital skills. Ensure that these opportunities are clearly communicated and supported across the organization.

(continued)

Takeaways for Chapter 9 (continued)

What We Know	What You Can Do about It
• Primary reasons employees gave for wanting to leave an organization were concerns with company viability and lack of opportunities to continue to enhance digitally relevant skills.	• Clearly communicate your strategic talent plan to employees. Make sure they know what skills will be needed so that they can help identify talent or develop those skills.
• Even digitally maturing companies reported needing more talent. They know that they possess a recruiting advantage and reach out regularly via LinkedIn and other platforms to cultivate desirable employees.	• Engage in passive recruiting to attract employees with valuable skill sets to more digitally mature areas of your organization. Passive recruiting can be conducted within your organization as well to place digitally savvy employees in pockets of the organization where they can thrive.
• Attracting sufficient digital talent requires assessing what skills will be needed in the future, locating to attract talent, and leveraging talent outside the industry or company.	• Recognize that there is a fast-emerging talent ecosystem of contractors, "gig" employees, crowdsourced work, and so on, and adjust your descriptions of the work and recruiting efforts accordingly.

10   The Future of Work

We all know someone who doesn't use email, doesn't carry a smart-phone, and shuns social media. Some of us might use the term Luddite to describe someone "who fears technology (or new technology), as they seem pleased with how things currently are."[1] As Richard Conniff writes for *Smithsonian*, "The word 'Luddite' is simultaneously a declaration of ineptitude and a badge of honor."[2] The term originated in conjunction with a British industrial protest more than two hundred years ago. The protests started on March 11, 1811, in Nottingham, a textile manufacturing center, at a time of economic upheaval, food shortages, and high unemployment. Disgruntled textile workers smashed machinery that night and subsequent evenings, inspiring a series of similar attacks across northern England. The workers naïvely thought that the destruction of the machines would protect their jobs. The government retaliated quickly, first by posting soldiers to protect the factories, then by passing laws that classified the destruction of machines as a capital offense.[3]

Today, we live in a different world, not least in economic outlook. Yet increasingly, we face uncertainty about the impact of new technologies on the work we do and on the organizations where we work. In some cases, this uncertainty is accompanied by fears of jobs being lost to cheaper and more effective technologies. As companies struggle to adapt to the changes wrought by technology over the past two decades, this disruption shows no signs of slowing down. If anything, the magnitude of the changes coming down the pipeline for the next decade or two will likely prove to be even more significant and disruptive. In this chapter, we consider the implications this ongoing disruption will have for individuals and organizations.

Some caveats are necessary, however, before we discuss how technology will disrupt work in the future. First, questions about *whether* these disruptions will happen are different from questions about *when* they will happen. Experts often disagree significantly about when certain digital disruptions will come to fruition, and many arguments waged by skeptics point out the difference between the current state of technology and the future promised state. Here, we focus more on the "if" than on the "when" disruption will happen to specific fields. We recognize, however, that a key aspect of strategic decisions is the need to understand when changes will happen and how quickly to respond to them. Nevertheless, comprehension of what types of changes are likely to happen is not without value. If managers have a good understanding of what disruptions are likely on the horizon, they will be better able to focus on signs or triggering events that indicate when particular changes will happen. Furthermore, understanding the types of disruptions that are coming can help people prepare for a changing future, regardless of when it finally arrives.

When technological disruption of human jobs happens, it will likely occur in two stages—first augmenting and enhancing the human worker, then replacing the human altogether. The implication of this perspective is that many jobs will become enhanced and improved by technology right before technology fully replaces them. Do not be fooled into thinking that disruption won't happen simply because technology makes human workers more valuable in the short term. This augmentation step will make the professionals more valuable in the medium term because it will free up experts to shift from routine to value-added tasks. The question will be whether human employees can develop these new value-added roles before technology takes over those roles entirely. Then, the question shifts to whether human employees can take on other value-added roles and work. Certainly, aspects of some jobs will never be fully replaced. For example, although some parts of a radiologist's job may be automated, we expect that most patients will prefer to receive a cancer diagnosis from a human not a computer. But just because certain aspects of a job have not yet been replaced by technology does not mean they can't or won't be.

The transition between augmentation and replacement will likely happen quickly when it does. The augmentation stage will probably be more protracted as people begin to get comfortable with automation supervised by humans. Yet, once people are comfortable with supervised automation, they may then rapidly decide that human oversight is no longer worth the cost or the extra effort. A similar dynamic occurred in the newspaper industry during the dot-com boom, when revenues slowly rose until they fell off a cliff shortly after 2000. The internet initially enabled established news companies to extend their reach and lower their production costs, thereby increasing advertising revenues. It was a boon for publishers until other competitors—such as Craigslist—began to enter the market and take a considerable portion of these revenues with much lower overhead. Today, Craigslist still has only about fifty employees and brings in around $700 million through classified ads, which used to be a key source of revenue for newspapers.

How Will Technology Disrupt Work?

In chapter 4, we discuss the potential effects of autonomous vehicles on several industries, including automotive, real estate, and insurance. These vehicles may have still bigger implications for work. Yet, even a disruption as significant as autonomous vehicles pales in comparison to the potential disruption in work by artificial intelligence (AI). Some estimates suggest that as many as 80 million jobs in the United States will be affected by AI, including those of telemarketers, paralegals, cashiers, fast food cooks, and various positions in the financial services industries.[4] AI is particularly well suited to replace either routine work or skilled jobs that are based on making predictions from past data. For example, radiologists spend years studying to distinguish the difference between normal and abnormal X-rays, CT scans, and other types of medical imaging, and they typically earn between $400,000 and $500,000 annually. AI will be able to train on millions of images in a matter of days, becoming far more accurate than its human counterparts. Even general managers might be at risk of being disrupted. The enterprise collaboration platform Slack is working on AI to monitor employee communications and automate many managerial tasks, reducing the need for face-to-face meetings.

Still other technologies are looming that will disrupt jobs even further. Although bitcoin is arguably the best-known application of blockchain, the potential impact of blockchain goes far beyond cryptocurrencies. As a technology that enables a secure public record, blockchain is poised to eliminate jobs that depend on mediating trust between parties. For example, it could be used to create self-executing contracts that would eliminate the need for escrow services. We can envision similar scenarios with additive manufacturing, virtual reality (VR), and augmented reality (AR).

Each of these coming technological trends alone—autonomous vehicles, AI, blockchain, additive manufacturing, and AR/VR—could have a significant effect on jobs over the next decade. Taken together, however, these multiple technological trends portend massive disruption in the future of work. Indeed, the path of digital disruption ahead suggests that we may be closer to the beginning than to the end of the type of disruptive influence that technology will have on work. In our experience, even though many people know this disruption is coming, employees and leaders are not generally considering how these technologies will affect their careers.

The Future of Work or the Work of the Future?

The impact of these technologies on work are complex and not entirely predictable. For example, as MIT economist David Autor notes, there are nearly twice as many bank teller jobs today as there were at the introduction of ATMs; but those teller jobs are very different than they were before.[5] They have become less about counting money and keeping records and more about developing relationships with customers and providing financial advice. Similarly, we can see how radiologists may spend less time discerning abnormal from normal images and more time focused on analyzing the abnormal images. Or managers can spend less time on task oversight and project management and more time on coaching, mentoring, and developing their teams. Autor notes that many pundits claim that "this time, disruption is different," but adds that people have always thought that the disruption they lived through was different from the previous ones. We can look back to see how previous generations adapted to deal with the disruptions they faced, but we can't look forward to see precisely how to adapt to the one we're facing.

Echoing this sentiment, Deloitte CEO Cathy Engelbert notes that she prefers to talk about these trends as the "work of the future" rather than the more commonly used term "the future of work." We concur with this shift in terminology, as we think the former terminology is far more optimistic (and, we believe, more accurate) than the latter. It implies a shift in how work will be performed in the future, rather than questioning whether work has a future and will still exist. Autor notes that, in many ways, full-time work doesn't "need" to exist now. If people were content with the standard of living as it was one hundred years ago, they would need to work only about seventeen weeks per year. Instead, people work harder and adapt their skill sets to improve their quality of life. Nevertheless, while we agree with the sentiment echoed in the formulation of "the work of the future," we have instead retained the more widely used formulation of "the future of work" because this terminology is the most commonly used to discuss these issues, in our experience.

In the past, people adapted their skill sets quite resiliently to the types of work that were available in the economy. In 1910, the most common job was farmer or farm worker, with farm-related employment making up nearly 40 percent of the US workforce. By 2000, only about 2 percent of US workers were employed in farming.[6] In contrast, around 20 percent of workers in 1910 were employed in professional, clerical, managerial, or service positions, but in 2000, about 70 percent of workers were employed in these categories. The skill sets of these workers have also shifted to adapt to these jobs. In 1940, less than 5 percent of workers had bachelor's degrees, whereas just over 33 percent have a bachelor's degree today.[7] This number was only 28 percent a decade ago, and the trend is driven by younger workers, 37 percent of whom have a four-year degree. As work has changed, people have adapted their skill sets to accommodate the demands of that work.

Of course, this shift to new ways of working will not be smooth and painless. It is likely inevitable that wide swaths of people will be unable to adapt and be left behind. A report by the *Atlantic* describes much of the societal disruption that accompanies these types of shifts.[8] People who are left behind because they are unable to adapt for whatever reason often experience psychological effects, societal disruption, and substance abuse. Of course, these societal effects of economic disruption are not "different this time," either. In Clay Shirky's book *Cognitive Surplus*, he argues that the gin craze of the 1700s was largely in response to the increasing urbanization and economic disruption in London. People drank to cope with the economic disruption they were living through. A similar reason might also undergird the recent opioid epidemic in the United States, as people struggle with the disruptions they are experiencing. The difficulties caused by shifts in work can likely only be addressed through public policy and government intervention, which are worthy topics but not ones that we address in this book.

We do think that, as in previous periods of technological disruption, workers and the economy will adapt to new demands. And, as in these previous periods, that process will often be painful and disruptive as people seek to adapt. This time may seem different because one can look back on previous examples and jump to the resolution without living through the uncertainty and difficulty required to get to it.

What Work Are People Best At?

Although we do not know precisely how people will adapt or what most jobs will look like, several pundits have pointed in the direction of identifying the areas in which people are better than computers. Some, like columnist and author Tom Friedman, suggest that providing caring is another way in which people are better than computers. He notes, "We used to work with our hands for many centuries; then we worked with our heads, and now we're going to have to work with our hearts, because there's one thing machines cannot, do not, and never will have, and that's a heart. I think we're going from hands to heads to hearts."[9] Anthony Goldbloom, the founder and CEO of Kaggle, suggests that making decisions from incomplete data is one way in which people are better.[10] This insight is related to something Pablo Picasso said of computers: "But they are useless. They can only give you answers."

While identifying the types of work that only humans can do may be a valuable exercise at times, it may not be the most productive way to prepare for the future of work. In theory, there are types of problems that computers are fundamentally incapable of addressing—such as Alan Turing's halting problem, in which he showed in a 1936 proof that a computer cannot tell whether it will successfully finish running a set of code prior to running it. In practice, computers have proved far better at performing tasks that we once thought impossible, like facial recognition and language translation. If we primarily fit human work into the gaps left by what computers cannot do, people will increasingly be squeezed out as technology becomes more advanced.

For example, Cynthia Brezeal, of MIT Media Lab, is designing so-called sociable robots that can approximate empathetic connections. Research has shown that people are also more likely to open themselves up to robots than to humans, because fear of judgment is significantly diminished.[11] So robots may, in fact, be capable of performing caring jobs in ways that people cannot. Simulations can also enable AI to make novel insights from past data that humans cannot. For example, when the AI system AlphaGo competed solely against itself to learn the game Go, instead of using data from human players, it was able to create insights and strategies that people working at the game had not developed over the centuries of playing it.[12]

This logically raises the question, How are people truly better than computers? As humans create and find new opportunities in responses to technological evolution, technology may evolve to take over those new roles eventually as well. Yet, this development will create even more new opportunities for work that humans can figure out. If, as Picasso suggested, people are good at asking questions, then what questions should we be asking? In the near term, one certainly might be, What are the new opportunities that arise as technology takes over certain aspects of work?

Seeking the Right New Opportunities

At first, autonomous vehicles will certainly give rise to different types of work. Doctors, nurses, lawyers, and other professionals may be more apt to conduct house calls, as they'll be able to use the travel time productively. People may be able to use their kitchens to start restaurants that rely on self-driving vans for food delivery. Certainly, still other new jobs are possible. Autor reminds us that not being able to envision them now doesn't mean they won't happen. The farmer who was disrupted in the 1900s probably did not envision the future job of data analyst predicting yield. We must not be ignorant to the fact that technology is likely to evolve to take over those new roles eventually as well—the sympathetic robot may one day replace the traveling human doctor. But we expect these changes to take place over time. Marco Iansiti and Karim R. Lakhani argue that it will likely be twenty years or more, for example, before blockchain becomes mainstream.[13] Even if technologies evolve more quickly, societies and institutions often change more slowly.

Before Picasso, Voltaire said, "Judge a person by their questions rather than their answers." Ironically, asking questions about new opportunities for work in light of technological disruption may, in fact, be the one task for which humans are inherently superior than computers. In many ways, the ability to ask these questions combines the earlier examples, nominated by others, of tasks that humans are inherently superior to computers at accomplishing. This ability to question is part Friedman's empathy, since it involves identifying unmet human needs and desires in the new environment. Questions are also part Goldbloom's decision making based on incomplete data, since the questions identify needs in a new environment created by technological evolution. In other words, the very task that computers may not be able to do better than humans is identifying opportunities created in the wake of technological evolution and disruption. Humans may be uniquely well suited to identify these gaps, adapting their skill sets and spending their time meeting these needs. Asking the right questions is, at least for now, a uniquely human capability.

Implications for Individuals: "Pivoting" on the Career Path

What are the implications of the work of the future for individuals? Perhaps most important, people need to prepare to be *lifelong learners*. As technology continues to change at an increasing rate, people will clearly need to learn new skills to remain relevant. This essentially requires an extension of the growth mindset throughout one's career. Identifying the ways in which humans can uniquely provide value to the human-computer partnership is one thing, but being able to execute on those opportunities is another altogether. People will need to develop new skills as they adapt to the changes wrought by advances in technology and the human-machine partnership.

Although one possible implication for this prediction is that people will need to continually learn new skills to remain in their chosen professions, a more likely interpretation of this dynamic is that the concept of a lifelong career will become an artifact of the past. The pace of technological disruption is such that any jobs people do at the beginning of their careers will be obsolete long before those careers end. Even if the jobs still exist, technology will have reshaped the work required to perform them to such a degree that the required skill sets will be almost entirely different. Instead, people will "pivot" to new careers as their skill sets become undervalued in one job or sector, requiring them to repurpose them in new roles or industries. This pivot may take the form of traditional retraining, or it may involve applying existing skills in new contexts, which, presumably, will provide workers with a new set of skills that would then be resources for the next pivot. Just as organizations need absorptive capacity to adapt to innovations, as we address in chapter 2, so ongoing learning and a growth mindset will allow individuals to remain flexible enough to develop new skills.

This need to pivot will mean that individuals will need to *chart their own career path* amid these changes in work. A metaphor for these types of career paths for the future can be found in surfing. Surfers catch a wave for a set period, riding it to its natural completion, at which point they must paddle out and look for the next wave to catch.[14] Some surfers choose to ride a wave as far as it will take them, while others choose to bail out once the wave passes its peak, so that they can be better positioned to catch the next wave. Likewise, some workers will choose to stay on particular paths longer, while others will attempt to pivot more quickly and jump from crest to crest. Regardless, organizations will likely need to help support these different career paths so that they can ensure access to the needed talent, in much the way Cigna charted and supported valued skill sets, as described in the previous chapter.

Allied Talent

Chip Joyce, of Allied Talent, envisions that companies will interact with employees in very different ways in the future. Drawing on work from Reid Hoffman's book *The Start-Up of You*, Joyce thinks that companies will engage with employees not through indefinite employment but through shorter tours of duty, designed to deepen skills while engaging employees in creating their career paths. A typical tour lasts for two to four years and is focused on specific goals that support both the corporate mission and the employee's career. Managers are committed to developing employee skills needed to complete the tour and to then discuss additional tours of duty based on both the company's needs and the employee's career goals.

Workers can choose tours of duty based on their current career goals. They may take an engagement at a slightly lower salary for an opportunity to develop new skills or one with more modest hourly work requirements when starting a family. Conversely, they could sign up for higher-paying engagements that maximize their existing skill sets and require working eighty hours per week when their career aspirations are so aligned.

Joyce envisions digital dashboards that allow companies to find employees with the right skill sets and the right career aspirations to match a given requirement. The result is not a one-size-fits-all approach to employment, but a more nuanced perspective that allows employers to match the right job with the right candidate given the current expectations of both. Joyce envisions this arrangement will help companies attract more motivated workers and enable employees to find the right opportunities given their career stage.

More than One Way to Pivot

Tom Davenport and Julia Kirby describe several different ways in which employees can pivot in their career path in response to digital disruption, except they refer to it in terms of five different "steps" employees can take.[15]

- **Step up.** When employees step up, they choose to develop the skills that will make them more valuable and marketable in a digitally disrupted business. Examples of this step include pursuing advanced degrees and continuing skill development to keep up with disruption. Companies could support this by developing a strategic talent development plan like Cigna's, described in chapter 9.

- **Step aside.** Employees who step aside develop strengths in areas that are not easily disrupted by technology, such as emotional IQ or tacit knowledge that isn't easily codified. An example here might be developing creative skills or tradecraft. This step may also help address the interest of companies in a combination of hard and soft skills.

- **Step in.** When someone chooses to step in, they begin to develop their skill set for the digitally disrupted industry. An example might be radiologists becoming adept at using and understanding computer diagnostics to monitor the diagnoses and learn when to intervene. Companies need to support employees' efforts to learn new technologies in their specialty.

- **Step narrowly.** In this situation, employees specialize deeply in an area that computers are not likely to disrupt in the near future. Davenport and Kirby use an example of a man who specializes in matching up buyers and sellers of Dunkin' Donuts franchises. It is a niche competency that may never attract enough attention for automation. Organizations may be well served to identify and support employees with these niche competencies, as they may become an important source of differentiation from competitors.

- **Step forward.** With stepping forward, workers attempt to get out ahead of digital disruption and develop the technology that will represent the next wave of disruption. These people would now be working on the next great application for blockchain or developing components for autonomous vehicles. Companies in various industries are supporting these efforts by funding and engaging with an ecosystem of startup companies.

The "Glass Is Half Full" Perspective

While it's tempting to mourn the loss of the security of a lifetime career, this destruction and creation of career paths does have some upside. We all probably know people who feel stuck in a job they dislike, simply because they feel that they cannot afford to pursue new opportunities. These dead-end jobs will be much less likely in the future of work, because the changes in technology will make linear career paths obsolete, and particular careers may not last long enough to become a dead end. Companies have already begun to adapt to these changes in individual career paths. Allied Talent (see above) suggests that companies adopt short-term tours of duty, with people placed in roles for a few years, at which point they are shifted to new roles. The upside of this approach is the prospect of continual learning embedded in the organizational structures and processes. People not only learn new skills in their new roles, but they also bring fresh perspectives and skill sets to these established jobs.

While some older workers may groan at the thought of needing to learn new skills late in their careers, we think this response primarily stems from thinking they wouldn't need to engage in continual learning. People coming into the workforce in the 1980s and 1990s thought they could engage in a set of skills for their entire careers, and they are understandably disappointed that they cannot do so. Learning these skills is also harder because they have not practiced lifelong learning. Workers of today will not share those assumptions, and they will be more accustomed and able to learn the skills as needed.

This need to continually pivot to the next possible career wave also has another implication—the need and/or the ability for employees to chart their own course of career exploration with *passion*. By passion, we don't necessarily mean an overriding and long-term desire for a specific goal. Instead, we envision it as the opportunity to scan the environment and find the point at which personal interest and market opportunity are maximized. The American writer Frederick Buechner describes this as one's calling, where the world's deep need and the individual's deep joy meet. The World Economic Forum describes this intersection in terms of the Japanese concept of *ikigai*—the junction at which what you love, what you are good at, what you can be paid for, and what the world needs all come together. We think these successive career waves can provide greater opportunities for employees to achieve ikigai, pursuing new avenues as their passions change and the disrupted world creates new opportunities to do so.

Conniff argues that the Luddites were (contrary to popular opinion) not opposed to technology per se. Rather, "the original Luddites would answer that we are human." In a world where people fear being replaced by machines, whether textile manufacturing machines, in the case of the Luddites, or robots and AI today, employees are looking for ways to find meaning from their contributions. It's not the technology that strips away meaning; it's the tacit assumption that the workers themselves are commodities that can be easily replaced. "Getting past the myth that people simply object to technology and seeing their protest more clearly is a reminder that it's possible to live well with technology—but only if we continually question the ways it shapes our lives."[16]

Takeaways for Chapter 10

What We Know	What You Can Do about It
• Technology will continue to disrupt all types of work, even if we do not know precisely how and when that disruption will happen. Paradoxically, work about to be disrupted may become particularly valuable before disappearing.	• Pay attention to how work is being disrupted and what types of skills are being replaced by technology. Identify the jobs in your organization that are likely to be disrupted in the next year, one to three years, and three-plus years.

(continued)

Takeaways for Chapter 10 (continued)

What We Know	What You Can Do about It
• For the work of the future, people will have to be lifelong learners, acquiring new skills to help them address the needs and opportunities created by digital disruption.	• For each job category that has the potential to be disrupted, create an action plan for how to handle the affected employees. • As appropriate, link these action plans to training and learning opportunities to ensure that employees have a chance to position themselves for the work of the future.

III Becoming a Digital Organization

11   Cultivating a Digital Environment

We live in a world in which most business leaders readily subscribe to the admonition attributed to management guru Peter Drucker: "Culture eats strategy for lunch."[1] Unsurprisingly, culture is important in almost every discussion of digital transformation. Organizations struggle with understanding the ingredients of a digital-first culture. Some companies make pilgrimages to Silicon Valley and even try to set up shop there in the hopes of breathing the same air and somehow absorbing the magic. Others try to create a digital-first vibe by designing sleek new office spaces with comfortable couches, open collaborative spaces, foosball tables, and a jeans-only dress code. Still others change the names of leadership roles to try to demonstrate that they're truly being digital, but they don't really know what it means beyond the name change (e.g., chief digital officer instead of chief marketing officer).

But digital culture runs deeper than that. It's not just about how an organization decorates its space and what tools it uses. It's about how an organization behaves, what it values, its unspoken but deeply embedded beliefs. Digital culture is often described as being "in the air" or part of the "vibe" of a place. Because culture feels nebulous, it's often viewed as "icing on the cake." As we discovered in our research, however, culture is *not optional* in the quest for success. In fact, we find that culture is a critical element of digital maturity, even if it is a bit hard to pin down. So, what *is* culture? It is often defined as the social behavior, norms, and beliefs of a group. It represents "the way things are done around here." Culture is not just what is written in a mission statement or a code of ethics; it is what people in the organization believe to be the accepted patterns of behavior. In this way, culture can be a strong enabler (or a huge hindrance) to digital maturity. In fact, in our study, an inflexible culture, complacency, and lack of agility are cited as the biggest threats companies face because of digital trends. In other words, an organization's culture can stunt or enable growth of its talent and leaders, as well as its overall digital growth and maturity.

Edgar Schein, Society of Sloan Fellows Professor of Management Emeritus at MIT, investigates organizational culture, process consultation, research process, career dynamics, and organizational learning and change.[2] Schein describes three levels of organizational culture.

- **Artifacts:** What we see. What a newcomer, visitor, or consultant would notice (e.g., dress, physical layout, furnishings, degree of formality).

- **Espoused values:** What they say. What we would be told is the reason things are the way they are and should be (e.g., company philosophy, norms, and justifications).

- **Underlying assumptions:** What they deeply believe in and act on. Unconscious taken-for-granted beliefs about the organization and its work, purpose, people, rewards, and so on.[3]

But, culture, particularly at the level of underlying assumptions, can be challenging to address. It is intangible, complex, and nuanced. Whole books have been written on creating the right kind of culture or environment in your organization. (A good one to start with is Ron Friedman's *The Best Place to Work*.)[4] In this chapter, we focus on three important points we've learned about digital culture:

1. Digital culture is critical to driving digital business adoption.

2. Digital culture is distinct and consistent, associated with digital maturity.

3. Digital culture is intentional.

Digital Culture Drives Adoption

One way to think about digital culture is to think about creating the right kind of *environment* within your organization—an environment that is necessary to get the most out of your people, your talent, and your leaders. As we discuss in part II, the success of an organization today requires that its people continually learn, adapt, innovate, create, and lead. The right talent and leaders will bring change and innovation—and, ultimately, growth to your organization. But, to get the most out of your talent and leaders, you need to build the right kind of culture and environment. In our research, we often heard of organizations that had hired digital talent or leaders who ended up struggling to make an impact because of the culture of the organization.

Culture can be compared to water in a fish tank. If you do not get the chemical balance of the water exactly right, your fish will die. As Wilhelm Johannsen, the evolutionary botanist mentioned in chapter 6, discovered with plant seeds, environmental factors can have a large effect on an organism's characteristics, growth, and ability to reach its potential.[5] In addition to helping get the most out of your people, culture is also an effective and important way to drive digital adoption and engagement in your organization. Companies at the three stages of digital maturity—early, developing, and maturing—have different approaches to leading change. While the difference between how early and developing companies accomplish this is nuanced, the difference between these and the most advanced maturing companies is far more striking (figure 11.1).

Early and developing companies *push* digital transformation through managerial directive or by technology provision. In contrast, maturing companies tend to *pull* digital transformation by cultivating conditions that are ripe for transformation to occur. This culture-driven, bottom-up approach is one we are actively exploring in our ongoing research. Our findings to date suggest that the top-down directive approach many companies are taking may be misguided.

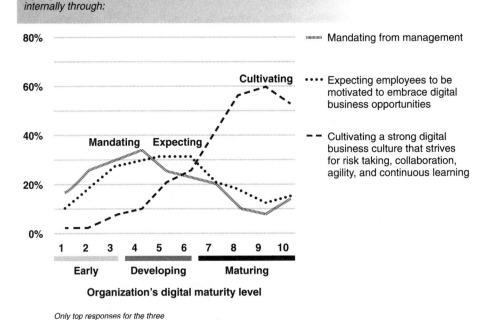

My organization primarily drives digital business adoption and engagement internally through:

- ⁗ Mandating from management
- •••• Expecting employees to be motivated to embrace digital business opportunities
- − − Cultivating a strong digital business culture that strives for risk taking, collaboration, agility, and continuous learning

Only top responses for the three maturity levels are shown.

Figure 11.1

Mandate from Management

Respondents from early stage companies reported that their primary method for driving digital adoption and engagement is to mandate initiatives from management. In this situation, organizational leadership decrees the nature of the next digital initiative, and employees are then expected to fall in line. A central problem with this approach is that top-down directives can often be surprisingly ineffective tools for driving adoption. The academic literature is replete with examples of employees finding various ways to avoid digital mandates when they want to, ranging from simply dragging their feet to actively sabotaging initiatives.[6] Employees can also use technology in unanticipated ways, which may or may not align with the mandate's business objectives. Even without these problems, mandating all the necessary behaviors to derive the desired business value from technology can be difficult.

Digital leadership requires approaches that differ from the entrenched command-and-control structures of traditional manufacturing age companies. President Harry Truman made a similar point about military approaches being ineffective in the US presidency when he said of General Dwight Eisenhower, his successor, "He'll sit here, and he'll say, 'Do this! Do that!' And nothing will happen. Poor Ike—it won't be a bit like the Army. He'll find it very frustrating."[7] Likewise, managers simply requiring that digital transformation happen is unlikely to yield the outcomes they hope to see.

Expect Employees to Adopt

Developing-stage companies follow a different approach. These companies expect employees to adopt digital platforms by building them— not dissimilar to the mantra "If you build it, they will come," from the 1989 Kevin Costner movie *Field of Dreams*.[8] While managers know that employees will not be driven by magical forces urging them to adopt new initiatives, they often don't provide the type of time, support, and motivation to adopt that would be necessary in other settings. Instead, companies often spend considerable time, money, and energy implementing digital platforms, expecting that the value of this technology will become so apparent to employees that they will be naturally drawn to use them to perform their work. And certainly, designing a user-friendly tool or platform with obvious value to the user is undoubtedly important. But companies that simply expect employees to adopt generally emphasize the technological side of implementation—and often execute that implementation well—while forgetting to accompany the new digital infrastructure with the organizational change management initiatives required.

Not only do employees need to be trained to use new technology, but they also need to be given time to figure out how to integrate these tools into their work. In research conducted with Lynn Wu of Wharton, we found that adoption of a new digital platform actually hinders employee performance for the first few months.[9] Only after about six months of use do organizations observe significant performance improvements.

Simply expecting employees to learn how to work with new technology while performing at pre-adoption levels puts employees and their organizations at a disadvantage for successful digital transformation. Such expectations are unrealistic, yet unfortunately quite common.

The German chemical company BASF provides a good example of how companies can provide employees the needed space to learn to use new technology. They encouraged project teams to adopt a new collaboration platform. At the same time, those teams banned the use of email for communication among members. The result was that the team needed to work together to figure out how to use the tool. Although they initially struggled with the new platform, they ultimately embraced it, with efficiency gains as a result of its features.[10]

Driving Transformation through Culture

Maturing companies drive digital transformation in an entirely different way, by focusing on creating environments where digital transformation can occur. Nearly 60 percent of respondents noted that their companies drive digital efforts by cultivating a strong culture that prizes risk taking, collaboration, agility, and continuous learning. Once the organizational conditions are ripe for digital transformation, leaders may discover that they have a much easier time engendering the types of strategic and technological changes that they need to compete. Once companies have cultivated an appropriate risk tolerance for the organizations, people are often far more willing to try new things. For example, although Google has gotten rid of its famed "20% time," during which people were encouraged to experiment, employees still retain a spirit of experimentation and risk tolerance in the culture which allows them to continue innovating.

Yochai Benkler, of Harvard, argues that employees are inclined to be more collaborative and cooperative, depending on the environmental conditions. He notes that in behavioral experiments involving the classic prisoner's dilemma game, 30 percent of people always cooperate, and 30 percent always act self-interestedly. The remaining 40 percent of people will decide based on signals from the environment regarding which approach is dominant. If they were told that they were playing the Wall Street game, this 40 percent acted according to rational self-interest, but if they were told that they were playing the community game, this 40 percent worked together and acted cooperatively. Sending the right cues to employees becomes an effective way of cultivating the right environment.[11]

As your company considers (or reconsiders) its own digital transformation initiatives, you should ask yourself whether you are approaching it in the right way. Are you pushing digital transformation on your organization, either through mandating adoption or by providing technology? Or, are you pulling transformation by cultivating the conditions that will elicit the types of change you desire? These differences may determine the ultimate success or failure of your digital transformation efforts.

Digital Culture Is Distinct

So, what kind of culture is needed to enable talent and leaders to help drive digital adoption and transformation? It turns out that digital cultures are like snowflakes—no two are exactly alike, particularly in what Schein describes as artifacts and espoused values. But just as snowflakes share a common set of distinct characteristics, such as their precise hexagonal array (or six-fold symmetry), digital cultures also share common and distinct traits. We asked a series of questions about a company's digital culture and then performed a statistical technique called "cluster analysis," which groups similar types of responses together.[12] Our cluster analysis yielded a clear finding—the culture of companies could be broken out into three distinct groups that were almost directly parallel to our early, developing, and maturing groupings. In other words, our data analysis showed that these types of companies had distinctly different cultures, even when our maturity data was not included in the analysis at all. This new analysis independently confirmed the three-tiered framework that we have been using for years and showed that companies at the different stages have distinct cultures.

To better understand the significance of these findings, however, note two important characteristics of this analysis. First, the maturity

Figure 11.2

ratings provided by the respondents were not included as a factor in the cluster analysis. The analysis returned our three digital maturity groupings, with a 90 percent correlation to our actual maturity groups, through an analysis of the organization's culture alone—irrespective of their digital efforts. In other words, our analysis showed that an organization's culture is inextricably related to its digital maturity.

Second, cluster analysis does not specify the number of clusters to return. It simply tries to find the number of groupings that fits best with the data. The cluster analysis could have just as easily showed four groupings or two or five. Instead, our results found that three maturity groups were the best way to explain similarities in organizational culture on the dimensions we identified. This analysis provided powerful, independent corroboration for our three maturity groupings (figure 11.2), confirming that they are the right way to think about the journey toward digital maturity—at least in terms of a distinctive culture.[13]

Our data show that a single set of cultural characteristics is associated with digital maturity, and these characteristics are consistent across industries and company size. Specifically, digitally mature organizations are:

- less hierarchical and more distributed in leadership structure;
- more collaborative and cross-functional;
- encouraging of experimentation and learning;
- more bold and exploratory, with a higher tolerance for risk; and
- more agile and quick to act.

These findings suggest that all organizations can begin the process of digital transformation in a single place, by working toward developing these hallmark characteristics of a digital culture. They provide further evidence for what we have already argued, that technology is only part of the story of digital transformation.

Digital Culture Is Intentional

The results of the cluster analysis may lead us to frame our theme even more strongly, suggesting that technology is, in fact, not even the most important part of digital transformation. If cultural characteristics are associated with digital maturity independent of an organization's efforts, and if digitally maturing organizations are driving digital transformation through cultural change, the most pressing challenge may really be more about shifting the organization's culture to be more adaptable to change. If companies can get the culture right, then resulting changes in technology and business processes can more easily follow. We should also raise an important caveat about these results. The responses used to categorize the company culture came from executives and employees alike, who were asked to describe how the company actually is—not what it talked about. The implication is that any attempt to make one's company more risk tolerant, more agile, with more distributed leadership, and so on must involve more than simple lip service from management on these issues.

As we note in chapter 3, many companies talk a good game about digital transformation. Just as they can talk extensively about digital strategy, so can they say all the right words about making their companies more agile and more risk tolerant. The companies that are actually able to bring about these changes, however, are far fewer in number. While the characteristics of effective digital culture are simple and clear, bringing them about is by no means easy. Yet, companies are succeeding in building cultures that are nimble, agile, collaborative, bold, and exploratory. How are they doing it? This leads us to the third important point we learned about digital culture: that it's *intentional*. Many digitally maturing companies make culture an intentional part of their efforts.

Salesforce's Intentional Culture

Salesforce grapples with the cultural challenges every startup faces when it grows—retaining the values and beliefs that were its essence at its founding. Salesforce preserves its digital culture through calculated efforts. "We're very intentional with our culture," says Jody Kohner, vice president of employee marketing and engagement. "Culture is not something that happens to us."

Being intentional about the culture starts with an emphasis on *'ohana*, the Hawaiian cultural value of extended family. "The concept is about an extended group of people that are bound together and responsible for each other," says Kohner. "We reinforce that sense of family from day one through actions, programs, and initiatives." Building trust and empowering career advancement are also intentional elements of Salesforce's culture. For example, to maintain and enhance its culture, the company asks employees for candid feedback and makes sure that they feel comfortable about being honest.

The trust the company builds translates to other values, such as empowering employees in their careers. "Silicon Valley is famous for people leaving for other companies," she says. "Our leaders and managers explicitly encourage employees to raise their hands when they want a new challenge. That helps us identify development opportunities and supports a culture of honesty, since there is no retribution for saying what you feel."

The Rich Get Richer

Our research shows that more advanced companies do, in fact, intentionally focus on culture as a critical part of their digital efforts. Perhaps more important, our data suggest that maturing companies are also making efforts to further develop the cultural characteristics that drive transformation. We asked respondents whether their companies are actively implementing initiatives to change its culture to be more collaborative, risk embracing, and agile in response to digital trends. These results show a strong relationship with digital maturity (figure 11.3). Of early stage company respondents, 23 percent said they are actively trying to develop a digital culture, and 54 percent of respondents from developing companies said so. For digitally maturing companies, however, a whopping 79 percent of respondents said their companies are

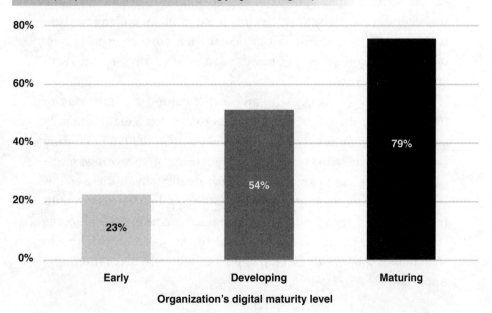

Figure 11.3

implementing these types of cultural initiatives, with the remaining 21 percent neither agreeing nor disagreeing.

In other words, companies that are already ahead in driving digital transformation are doubling down on these efforts to move their companies even further down this road. The digitally rich will keep getting richer. The companies that are already the most collaborative, agile, and risk tolerant are the ones that are also most likely to strive to become more so. Some of our interview data suggest that once these companies experience the benefits of these cultural changes, they want to keep moving in that direction. They also recognize that the natural tendency of organizations is often to move toward a stable state that becomes resistant to change, so they must constantly work to maintain the necessary flexibility for ongoing transformation. Other companies recognize that as they grow and continue to hire and assimilate new employees from other cultures, the need to maintain their own digitally mature culture becomes even more important. At the same time, they are likely facing the challenge of preserving and strengthening their culture across a larger and expanding organization. Regardless of the reason, the approach is the same. Only by continually working on these hallmarks of effective culture can digitally maturing companies keep driving the ongoing digital transformation necessary to keep pace with an ever-changing world.

Conversely, companies who are further behind are less likely to be considering these types of initiatives to develop a digital friendly culture. They may simply be less motivated than digitally maturing companies, or more likely, they may be focusing on the wrong aspects of the organization in an attempt to mature digitally. If these trends continue, we are likely to see a widening gap between the maturing and early stage companies, as the leaders continue to redouble their efforts to mature further. Because of the speed at which technology is changing, these characteristics are only becoming more important.

Slack—Building Culture in a Fast-Moving Environment

Slack Technologies, the messaging software platform based in San Francisco, doubles down on building a culture for digital maturity to make sure it can keep pace. In a fast-moving environment, company culture is what Slack's chief marketing officer, Bill Macaitis, says keeps him up at night. To fortify the company's culture, Macaitis relies on three principles:

- **Hire for it.** Slack's hiring process screens for all its core values. "When we hire people, we look at how empathic and courteous they are, along with evidence of their craftsmanship," he says. "When we make an offer, we are confident that the candidate aligns with our values."

- **Rally around it.** Slack reinforces the culture of empathy by embracing an "everyone does support" ethos. Designers, developers, and product managers work alongside customer support agents to answer support tickets. This helps build empathy as the people building the product hear firsthand about the problems existing customers have.

- **Live it.** "We've written our values on the walls," says Macaitis. "But as the old saying goes, it's not just what you write, it's what you do. You have to lead by example and provide ample training. That is what we are trying to do."

"One Weird Trick" for Digital Transformation

The internet is replete with websites purporting "one weird trick" that is all you need to accomplish weight loss, obtain a lifetime of wealth, or attract that perfect date.[14] While most of these promises are too good to be true, the results of our research suggest that "one weird trick" of developing a robust culture may be a critical factor for successful digital transformation. At the least, it provides managers with a clear place to begin working on digital transformation. Even if you do not know the right platform or strategy for your company, you can begin working on cultivating these cultural features.

For companies struggling with identifying which parts of their business to make more digitally mature first, developing a more digital culture seems like a compelling place where all companies can start. There is one clear path to digital maturity that all companies can follow across company size and industry—at least in terms of effective culture.

It provides a powerful road map for executives trying to increase the digital maturity of their companies. In the next three chapters, we dig deeper into specific cultural traits, particularly agility, collaboration, risk taking, and experimentation.

Takeaways for Chapter 11

What We Know	What You Can Do about It
• Companies at different digital maturity stages drive digital transformation in different ways. Early and developing-stage companies tend to rely on "push" techniques, specifically mandates or technologies being provided top down. Existing research suggests that these techniques can be ineffective in achieving digital transformation. • Digitally maturing companies, in contrast, tend to "pull" digital transformation by cultivating an organizational culture that creates the conditions that facilitate transformation. • Digital cultures are distinct and intentional.	• Using the organizational culture characteristics identified in this book, complete an assessment of your current organizational culture. • Compare where you are with where you want or need to be along different digital culture dimensions (e.g., risk taking, agility, collaboration). • Examine your current digital initiatives from an adoption standpoint to identify challenges that may be linked to culture. • Determine if you are using a push or pull approach for enabling these initiatives. Where you are pushing, determine how you can shift to a pull approach.

12 Organizing for Agility

In early 2001, seventeen software developers gathered in Snowbird, Utah, to discuss their shared ideas and various approaches to software development. The result of their discussion was the Manifesto for Agile Software Development, built around four values—individuals and interactions over processes and tools, working software over comprehensive documentation, customer collaboration over contract negotiation, and responding to change over following a plan. These values, supported by twelve principles (e.g., Principle 6 is that the most efficient and effective method of conveying information to and within a development team is face-to-face conversation), are the foundation of the Manifesto for Agile Software Development.[1]

"Agile" describes an approach to software development that substitutes rapid, iterative sprints for the more traditional "waterfall" approach, which moves sequentially through several distinct phases—requirements, analysis, design, coding, testing, and operations. A key problem with the waterfall method is that if the requirements are not well understood in the beginning, the resulting product might not meet the intended user's needs. The method is not well suited to the changing environments that characterize today's world. Agile concepts are no longer confined to software development. As we note in chapter 1, respondents reported that their organizations are too slow to change, too complacent, and don't have a sufficiently flexible culture to adapt quickly enough to changes in the competitive environment wrought by technology.

The solution to these problems for many organizations lies in methods derived from agile software development (colloquially referred to simply as the noun "agile"). Agile has "move[d] into mainstream management thinking, with some observers proclaiming it the next big thing."[2] As *Forbes* contributor Steve Denning writes, "Agile's emergence as a huge global movement extending beyond software is driven by the discovery that the only way for organizations to cope with today's turbulent customer-driven marketplace is to become Agile. Agile enables organizations to master continuous change. It permits firms to flourish in a world that is increasingly volatile, uncertain, complex and ambiguous."[3]

Principles of Agile Development

Many principles and practices of agile software development are applicable to the types of changes and struggles that companies in all industries are dealing with to mature digitally. Agile methodologies, however, attempt to increase variability in the outputs, which is beneficial when the "best" approach is not always clearly known. These methods are valuable approaches for cross-functional teams to work. Agile methods forego the careful planning of more traditional development methodologies, adopting a more test-and-learn, iterative approach to development. The goal is to get to a minimum viable product (MVP), which can then be iterated with the customer in a cycle that brings continuous improvements with each subsequent, rapid release. Agile software proponents emphasize several key principles of the development process.

The first set of principles deals with collaboration and communication among key actors. The focus on individuals and interactions, as well as customer collaboration, is grounded in the need for strong communication among all stakeholders in the development projects. These open lines of communication help identify and articulate when the processes diverge from what is expected or desired. This divergence could result from a misunderstanding of the requirements, from changes in the environment, or from other factors. We address collaboration in further detail in chapter 13.

The second set of principles is focused more on the process of developing the product. Delivering working software and responding to change occur hand in hand with iteratively developing the product. The team develops a working software product, and users figure out its key strengths, weaknesses, and missing features, which becomes feedback for the next iteration of development. Each iteration can be thought of as an experiment that tests whether the next version of the product is closer to the desired goal. Chapter 14 deals in greater depth with how companies can use experimentation and iteration to drive change across the organization.

Agile principles are an effective approach to digital transformation in organizations. Instead of planning a long road map of how to respond, teams develop short-term initiatives to engender small-scale changes in the company and in its processes. Next, the team evaluates how well the intervention meets its intended goals, and then plots the next small intervention. Agile teams don't develop a grand plan for digital transformation; rather, they take one small action at a time, assess its effect, and do it again. Management must effectively communicate to these teams about the strategic direction of the organization, as well as pay attention to and circulate the results of the teams, so that they come together into a set of meaningful changes. Instead of directly telling the teams what next steps to take, a process of ongoing two-way communication permits action above planning, and future action based on present results.

Adapting Your Organization to a Changing Digital Infrastructure

We are not interested in agile methods for their own sake. We are interested in them to the extent to which they can allow organizations to become more adaptable to a digital environment—sensing and responding more quickly. Respondents to our surveys describe digitally maturing companies as able to act rapidly. To differentiate this process from agile methods, we refer to it as *strategic agility*—an organization's ability to adapt to a changing market environment that occurs as the result of new or evolving technological developments. Because digital threats and varying rates of change are unpredictable, companies must constantly scan the environment for strategic threats and opportunities posed by technology. For example, few imagined that the ubiquitous adoption of smartphones would lead to a competitive challenge to the taxi industry through the rise of Uber.

Some of these changes to industry structure take many years, while others seem to happen in the blink of an eye. As former secretary of state Condoleezza Rice has observed, the timeframe and inevitability of change are much easier to understand retrospectively than prospectively. Responding too quickly and aggressively to changes may leave companies adapting to the emerging technologies prematurely and potentially squandering time and resources that might be best utilized responding to more imminent threats. On the other hand, failing to understand the urgency of a threat can leave companies in the dust and unable to recover. Although the impact of additive manufacturing (a.k.a. 3-D printing) has yet to be felt by many industries, it has completely disrupted the hearing aid industry in a matter of months.[4] Additive manufacturing of hearing aids uses laser scanning to map the patient's ear, producing a customized product more quickly and with higher quality than traditional methods.[5] Companies that were quick to respond to and capitalize on these changes thrived, while companies that responded too slowly did not.

The Digital Advantage of Cross-Functional Teams

Understanding which threats and opportunities your organization should respond to and which it should not is a difficult challenge to address through traditional hierarchical structures. A top-down decision-making culture puts considerable pressure on managers to get these strategic decisions right time and time again—a difficult, if not impossible, challenge. Instead, a more effective approach may be cultivating bottom-up decision making that relies on cross-functional teams.

If set up appropriately, cross-functional teams can have three key strategic advantages over traditional hierarchical and bureaucratic organizations.

• Cross-functional teams can act more quickly than a bureaucratic organization can, making decisions without the lengthy approval or socialization processes that are common in larger companies. Because these teams are made up of people from different functions within the organization, communication and socialization are faster and more continuous across the company. And within reason, these teams can act as they see fit in response to a digital threat. Teams often share knowledge with one another through "open house" meetings, where teams share their results and challenges with other teams.

• Different teams can tackle separate initiatives and pursue various options simultaneously. Managers do not need to decide which digital threat to respond to; they can simply determine which of the available options for responding should be exercised at a particular time. Teams that are tackling more pressing strategic threats can be provided with more resources, while teams that are dealing only with potential threats can continue to explore options to leverage when needed.

• Cross-functional teams also encourage employees to think differently. Because the teams are made up of people working in different disciplines, they bring in various perspectives and expertise to tackle a common challenge. "People have been focused on business capability delivery but just within particular segments of the business," says Freddie Mac's Christine Halberstadt. "Until you have a cross-functional view, you can't ask people to think differently."

Approximately 80 percent of digitally maturing businesses use cross-functional teams to organize work and implement digital business priorities, compared with around 20–30 percent of early stage companies (figure 12.1). In addition, respondents from companies that do so are far less likely to say that organizational structures are a barrier to success. Cross-functional teams seem to help organizations overcome one of the biggest barriers to digital transformation that companies face. Many people are

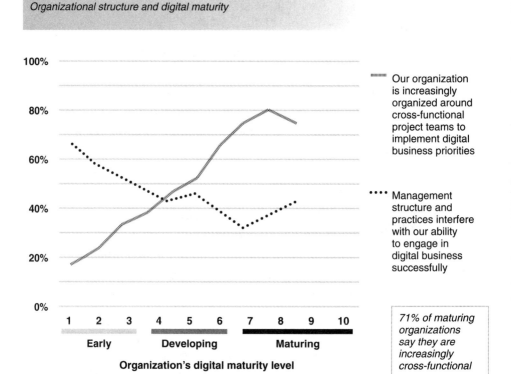

Organizational structure and digital maturity

Our organization is increasingly organized around cross-functional project teams to implement digital business priorities

Management structure and practices interfere with our ability to engage in digital business successfully

Organization's digital maturity level

Early Developing Maturing

71% of maturing organizations say they are increasingly cross-functional

Figure 12.1

familiar with the Amazon two pizza rule, formulated by Amazon CEO Jeff Bezos—if it takes more than two pizzas to feed the team, it's too big.[6]

The auto retailer CarMax relies heavily on cross-functional teams. CIO Shamim Mohammad says that teams "are empowered since the leadership team never tells them *how* to solve a problem, but *what* the problem is and the KPIs [key performance indicators] to work against." This approach allows for increased feedback, a significantly faster pace of development, and trial and error to ultimately arrive at a solution that is best for customers and associates. CarMax has also found that teams take smarter risks and are more creative in how they meet their objectives. If you don't have these fully integrated teams, then take a hard look and consider if you're set up for a successful digital transformation.

Interviewees provided several different reasons for the use of cross-functional teams in digitally maturing organizations. To some degree, the drive toward such teams is also inherent in technology that changes how work is done. Teams cannot explore strategic options for the organization if their members are employees from only one functional area. "It's just more difficult to think about any function in isolation because the processes are becoming so integrated," says Dave Cotteleer, of Harley-Davidson. "The opportunity for integration and collaboration is so great that it drives greater effectiveness and efficiency." As an example, Cotteleer points out that connected vehicles demand a stringent cross-functional approach to design and manufacture. "It's no longer just about product engineering," he says. "It is about software design, system integration, and other elements that fall outside traditional product engineering. Multiple functions in the company are now realizing that what used to be their domain is now also a domain of technology."

Cross-Functional Hospitality at Marriott

When George Corbin, Marriott's former senior vice president of digital, tried some competitor apps, he found that although they worked well technically, they didn't deliver—he wasn't checked in when he arrived, and he never received the dinner he ordered.

The experience sparked an important realization: "We can create the best website on the planet and we can build the best search campaigns to reach customers," he says. "But if we can't deliver an exceptional stay, guests won't come back."

To tackle the guest experience head on, Corbin began working closely with his counterparts in operations and, as he recalls, spent more time with them than with his own team. Corbin made good use of the operational knowledge and has been able to mobilize ground forces to make sure that Marriott's apps deliver on their promises.

Cross-functional teaming has become a permanent fixture. Multiple functions at the global hotelier now have the same performance metrics, including operational effectiveness and costs. There are also digital professionals in almost every function, not just in the digital units. "We all work with the same scorecard and address issues together," he says. "The company takes it seriously. A number of our metrics are reported all the way up to the CEO."

Empower Teams to Act

The nineteenth-century Prussian field marshal Helmuth von Moltke the Elder noted, "No plan of operation will be more than enough for the first encounter with the enemy's main force."[7] Likewise, agility means that teams are able to adapt their approach to a changing environment. Cross-functional teams are not likely to achieve agility unless they have some degree of autonomy. While it is critical for senior management to keep its finger on the pulse of the competitive environment, these cross-functional teams must be given a certain degree of autonomy to adapt their plans to the environment.

Part of the reason some digital teams do not act strategically is that they are not permitted by their companies to do so. Many organizations' strategies are locked up in the board room and the c-suite, with the average employee unaware of—or unable to act on—the bigger strategic vision. Instead, digitally maturing companies are working at pushing decision making to lower levels of the organization. Julian Birkinshaw, in his *MIT SMR* article "What to Expect from Agile," describes five lessons learned from ING Group's agile journey.[8]

1. Decide how much you are willing to give up. Agile shifts power from executives to others in the organization, which can be difficult. Executives must decide how much control they are willing to cede.

2. Prepare stakeholders for the leap. Agile is a different way of working, and you must prepare people for the change.

3. Build the structure around customers—and keep it fluid. This shift does not just focus on customer needs but reorganizes the company around them.

4. Give employees the right balance of oversight and autonomy. The need for top-level oversight doesn't disappear—it just changes. Figuring out the right balance may take trial and error.

5. Provide employees with development and growth opportunities. One risk to agile is that employees become too task-oriented and don't develop their skills. Effective mentoring is essential to keep employees developing.

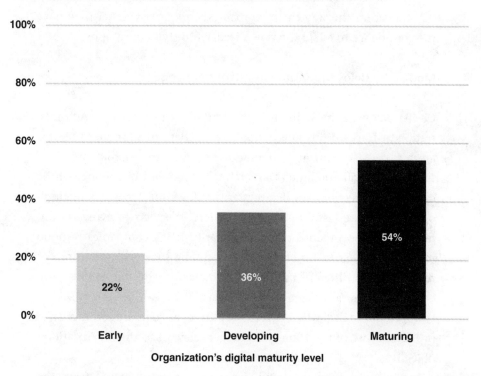

Figure 12.2

When we asked respondents if their organizations push decision-making authority down to lower levels of the company, their replies illustrated strong difference between digitally maturing companies and their less mature counterparts. Fifty-four percent of respondents from digitally maturing companies agree that their organizations are doing this, while only 20 percent disagree. In contrast, only 22 percent of respondents from early stage companies agree, while 54 percent disagree with the statement (figure 12.2).

Managing in this environment requires allowing, enabling, and empowering employees to act strategically and with less direct oversight. It also necessitates educating these employees on the organization's strategic objectives, as well as developing "rules of engagement" to guide employee action. Finally, it requires that executives learn new ways of managing and leading. The most pressing questions facing organizations are whether executives have the courage to let technology transform the way their companies work and whether they have the confidence to adapt their own leadership styles in response.

Modularity Helps Organizations Adapt to Change

Carliss Baldwin, the William L. White Professor of Business Administration at Harvard Business School, notes that cross-functional teams are a type of organizational modularity. She argues that organizations are actually designed to match the dominant technology of the day.[9] Just as technological modularity allows for computer technology to innovate faster by allowing different aspects to evolve at different rates, so organizational modularity helps companies respond more quickly. Cross-functional teams can be swapped in and out or repurposed to different tasks as needed. Teams work on specific projects, with a certain degree of autonomy. They can be repurposed or swapped out to deal with different objectives as goals are accomplished or the purpose of the team is no longer relevant to the organization's objectives.

Modular organizations will likely look quite different from the so-called flat organizations, with relatively few levels of hierarchy, that were heralded at the dawn of the internet era. Organizational theorist Karl E. Weick differentiates between tightly and loosely coupled organizations.[10] Tightly coupled organizations are legacy companies that have a clear set of policies and procedures reinforced through various feedback and reward mechanisms. In contrast, loosely coupled institutions push decision-making capabilities down through their hierarchy to better deal with conditions on the ground. Executive leaders communicate the strategic objectives of the organization, and the frontline leaders are partially responsible for figuring out how to achieve those objectives. These organizations remain deeply hierarchical, but these hierarchies operate differently than modern corporations.

Cultivating Talent Markets

A key aspect of modularity is the ability to source talent quickly and reliably as needed.[11] Companies may need to consider how to manage specialized, fluid talent differently than they manage traditional employees. In chapter 1, we note that Melissa Valentine, of Stanford, envisioned companies organized around a core-periphery model, consisting of a small group of core employees assembling teams from a peripheral group of temporary employees. One executive we interviewed notes, "Organizations will become a lot more fluid. The degree of ambiguity will be increased, and the degree of speed required will be increased. They need leaders who are able to assemble at any point in time a coalition of people who are guided by purpose, more than task or functional area necessarily."

The first step may require that organizations partner with or cultivate on-demand talent markets that sources and integrates talent across networks so that specialized help is available as needed and on demand. Talent markets can be maintained via platforms that monitor, evaluate, and support the talent pool of on-demand contractors. Companies that are loath to assume that people with the needed talents are available in the broader marketplace may instead develop and manage on-demand talent markets comprising both employees and freelancers. For companies to ensure access to the types of skill sets they need, they should recognize on-demand talent markets as strategic resources and invest in the long-term health of the talent pool. Individuals may come and go, but the on-demand talent market should be nurtured and maintained with an eye toward the future.

Some members of this talent market may remain full-time employees of the company, while others may be part-time employees or contract workers. While talent markets have typically been used to manage part-time freelancers, some companies have also begun experimenting with these markets as platforms for assigning full-time employees to projects as needed. For example, Work Market (recently acquired by ADP), has set up dedicated talent pools of full-time employees and part-time freelancers for individual companies. Full-time employees provide a stable base of employees, while part-time contractors offer the flexibility to deal with variations in demand. For some on-demand contractors, the opportunity to become full-time employees may be a powerful motivator to continue building their skill sets. Many people who are not in the market for full-time employment, however, still have valuable skills (for example, student workers, parents of young children, and people at or near retirement age). The crowdsourcing site Innocentive, based in Waltham, Massachusetts, has found that retired workers with specialized expertise are among its most valuable and regular contributors.

Many organizations treat contractors as second-class citizens, but companies that want to attract great talent can't afford to do that. On-demand talent with valuable skills can choose to work for any project or company. To ensure that they're able to get the best, organizations should cultivate an environment and incentive structure where on-demand contractors are valued as integral contributors to the company's strategic objectives. Providing desirable work experiences and environments, opportunities to work on interesting projects, and exposure to different teams can help drive engagement.

Rethinking Core Employees

Companies that increasingly rely on these talent markets may also need to rethink the nature and roles of the employees who will assemble and lead these modular teams. Core employees are not just full-time employees. They are the people companies will invest in to build and guide the long-term strategic direction of the organization. Although core employees will likely be working with other core employees, increasingly they may be delegating work to on-demand talent, which will require specific managerial skills. Effective delegation requires knowing how to source critical skills, how to assemble teams and get them up and running quickly, and how to use decision support tools effectively to meet the goals. These skills can provide the organizational agility and the collaborative environment that characterizes digitally maturing companies.

Core employees, even those who are relatively junior, should have a certain level of strategic autonomy to accomplish or contribute to designated goals. Strategic thinking is one skill that respondents to our survey listed as essential for both leaders and employees, and distributed leadership is a key cultural element of digitally maturing companies. Obviously, offering greater independence would require more communication with top leadership and increased awareness of the strategic direction of the company.

It is no accident that a key differentiator of digitally maturing companies is the way they intentionally work to develop, maintain, and strengthen employee engagement. Keeping core employees engaged for the long term involves providing more than a paycheck. For employees to want to stay and contribute, many say they need to feel that the organization is willing to invest in them and will continue to offer opportunities for growth. The 3M Company, for example, invests in new hires to build loyalty. According to 3M CEO Inge Thulin, the company plans to put all employees in an expanded employee development program by 2025. Core employees likely require new opportunities to grow their skill sets over time. Companies can create new development programs—unlike traditional leadership development programs that chosen employees take part in at certain points of their tenure—that encourage core employees to update their skills continuously to stay abreast of the ever-evolving world.

Flash Teams: Periphery without a Core?

The emphasis on multifunctional teams and bottom-up leadership may seem like the end of specialization, but Stanford University's Melissa Valentine doesn't think that has to be the case. "People can become valuable as they specialize," she says, and it can also afford them new opportunities to develop. As assistant professor of management science and engineering, Valentine's work focuses on crowdsourcing and a relatively new concept known as flash teams—adaptable, flexible alternatives to traditional work flows and roles.

Flash teams are computationally guided teams of crowd experts, supported by lightweight, reproducible, and scalable team structures. They are established through a web platform that gathers workers and manages them as they follow a structured work flow defining each task and how workers interact. For example, "A back-end developer who shows up at a flash team meeting can do data structure work without a lot of supervision," she says. Their specific expertise can add "a very strong sense of what that kind of development looks like."

As long as people don't get locked into one niche, they can use their background to create new business value, according to Valentine. Flash teams

(continued)

Flash Teams: Periphery without a Core? (continued)

> thrive on their openness to new outcomes and their spontaneity. They work especially well for innovative projects, "where you're prototyping, piloting, or building something new," she says.

Different Ways of Working

As Birkinshaw and others point out, agile is not just about rapid, iterative approaches. It's also about different ways of working. His article concludes with the following observation. "ING's experiences are a reminder that implementing new practices is much more difficult than suggesting them. . . . No wonder new management practices often work better at young companies than they do at old ones, where the employees [and leaders, we would emphatically add] have entrenched expectations and habits."[12]

Takeaways from Chapter 12

What We Know	What You Can Do about It

• Companies need to be agile to respond to the changes and speed of the digital environment.
• Companies can look at the values and principles of agile software development methods as an approach to digital transformation:
  ◦ Individuals and interactions over processes and tools
  ◦ Working software over comprehensive documentation
  ◦ Customer collaboration over contract negotiation
  ◦ Responding to change over following a plan
• Agile approaches emphasize small meaningful improvements involving sprints over short (six-to-eight-week) timeframes and constant face-to-face communications over formal project reporting.
• Digitally maturing organizations tend to rely on cross-functional teams that are empowered to act and operate in an environment of reduced bureaucracy.
	• Conduct an audit of how you currently run your digital transformation initiatives relative to the values and principles of agile. Identify the biggest gaps.
	• Select one or two existing or planned digital initiatives that can be run as agile efforts. Train key members in agile methodology and run a learning pilot focused on understanding what it will take to scale agile more broadly. Determine what additional infrastructure and process changes will be needed for the agile pilot.
	• Evaluate the pilot and repeat.

13 Strength, Balance, Courage, and Common Sense: Becoming Intentionally Collaborative

Imagine a competitive team-building exercise in which you and your colleagues are told to build a tower by climbing onto one another's shoulders until you've reached ten levels, all standing on each other's shoulders. The winning team is the one that assembles and disassembles its multilevel human tower most quickly. If you have ever been to Catalonia, you may have witnessed exactly such a competition—"a trembling tower of red clad bodies [that] rises tier by tier, the broad-backed men at the base sweating and shaking under its weight, until the tiny girl shimmies to the summit and raises her arm in victory. It's a terrifying spectacle, but one of which its fearless participants, or *castellers*, are tremendously proud, for *castell* building is central to Catalan culture."[1]

The Catalan tradition of building human towers, or *castells* (the Catalan word for "castle"), dates to the eighteenth century. "The *castell* is considered successful when all the members manage to climb onto their designated positions and the *enxaneta* (the topmost person, usually a child) raises their hand with four fingers erect, which is a gesture symbolising the stripes of the Catalan flag."[2] The tradition gained popularity in the 1980s, culminating in its addition to UNESCO's "Representative List of the Intangible Cultural Heritage of Humanity" in 2010.[3] UNESCO describes the process of building a castell and how the "knowledge required for raising 'castells' is traditionally passed down from generation to generation within a group, and can only be learned by practice."[4]

If you want to see real-life castellers in action, just search for the term on YouTube, which has dozens of videos available.[5] Disassembling the tower without incident looks even more formidable than assembling it. The motto of castellers is "Força, equilibri, valor i seny" (Strength, balance, courage and common sense).[6] In a world disrupted by technology, the Catalan tradition of castells and its motto may be the perfect metaphor for the types of collaborators and collaboration required by digitally maturing organizations.

As we discuss in the previous chapter on agility, digitally maturing organizations are less hierarchical and increasingly organized around cross-functional teams, and they drive more decision making down into lower levels of the company, where those decisions can be made more quickly and in a more informed way. Tying these components together is a need to be more collaborative. As we note in chapter 11, one of the distinguishing cultural characteristics of digitally maturing organizations is that they *are* more collaborative—and *intentionally* so. While only about 30 percent of respondents at early stage companies agreed that their company is collaborative, nearly 90 percent of respondents from digitally maturing companies did.

What Drives the Need for Collaboration?

The reasons that respondents provided for what is driving their collaborative efforts may be even more intriguing. While we expect that the nature of work is the main driving force behind increased collaboration, respondents reported that *both* the nature of work and the availability of new tools and technologies for collaboration are drivers. In other words, people collaborate in new and different ways because the work demands it *and* because they now have the tools that enable them to do so more effectively. Maturing companies put these sentiments into practice and are considerably more likely to adopt more advanced collaborative tools (as opposed to relying primarily on email). While over 70 percent of respondents from digitally maturing companies reported that they are working or starting to work with advanced collaborative tools, less than 40 percent of respondents at early stage companies did (figure 13.1).

The nature of work in a digital age requires organizations to work more across functions, become more agile, and operate more iteratively. More collaboration is the obvious way to respond. When used effectively, digital

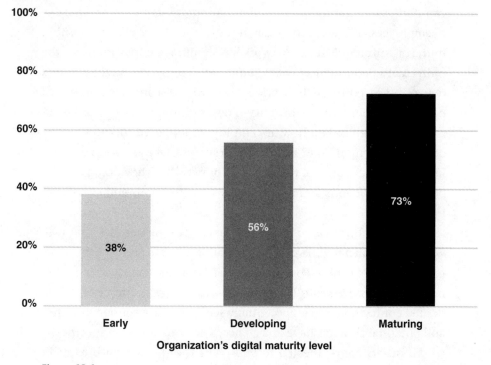

My department or team is working or starting to work with advanced
collaborative tools instead of email to facilitate better communication
(Respondents who answered "strongly agree" or "agree")

Early   Developing   Maturing

Organization's digital maturity level

Figure 13.1

platforms not only enable collaboration, but also change the way people
can collaborate and relate to one another. Given the potential collaborative benefits of more advanced platforms, it is somewhat surprising how
slowly organizations have adopted these tools for internal communication. After all, most individuals in the developing world began using Facebook and LinkedIn five to ten years ago, but many of the collaboration
features embedded in those technologies are just now beginning to make
it into the enterprise through platforms like Slack, Jive, and Salesforce.

Digital collaboration tools can help groups communicate vastly more
efficiently and effectively. Specifically, platforms provide two key capabilities in the enterprise context that support better and more intentional collaboration—managing networks and sharing content.

Managing Social Networks

In terms of *managing networks*, social media platforms provide new capabilities for relating to others that have previously been unavailable to employees. A long line of research shows that an employee's position in the organization's social network has significant implications for the individual's performance. For example, occupying structural holes in a network (i.e., positions that bridge otherwise disconnected groups of people) has informational benefits in terms of timing, access, and referral of information. Yet, how are employees supposed to know if they occupy a structural hole? Research also shows that most people have a limited understanding of the structure of networks they occupy; interpersonal interactions and relationships are amorphous and sometimes hidden.

Digital platforms can offer transparency into the network that real-world and email networks do not. They allow users to visualize the social relationships of others. For example, Facebook automatically identifies mutual friends. LinkedIn lets you know the shortest and best path to desired contacts. Users can potentially see how many connections they and potential contacts have at a glance. They can also note specifically whom others are connected to and chart the shortest path to make desired connections. By being aware of the characteristics of the larger social network, they can make better decisions about how to improve their place within it and, in turn, gain access to the performance benefits associated with those positions.

Managers can also use this transparency provided by digital platforms to gain an overview of the enterprise's entire network, showing which employees interact with one another.[7] Such a perspective can provide managers invaluable insights into how the organization is functioning. The 1982 best seller *In Search of Excellence* popularized Bill Hewlett and David Packard's practice of "management by walking around" mentioned in chapter 9. Network visualization tools can allow managers to "walk around" without ever leaving their desks—though we believe it's important to do both. These tools can provide a powerful lens into how the organization functions, how it changes over time, and whether interventions to improve that network structure are effective.

"Sociometrics" is a term developed by psychotherapist Jacob Moreno to describe the study and measurement of relationships and social structures, as well as the measurement of groups and the behavior and status of individuals within such groups.[8] For example, the analytics firm Humanyze helps companies utilize sociometric data generated by employees through existing digital platforms or custom employee badges to learn about employee behavior and make changes to the organization. Managers can monitor changes in communication patterns to determine if those changes had the desired effect. For example, Humanyze learned that changing the size of lunch tables in the cafeteria led to more productive discussions among employees, and synchronizing employee breaks had a similar effect.

Humanyze CEO Ben Waber describes four different levels of sociometric capabilities now available though still in early stages of rollout:

- First, you analyze the effects of a decision you have already made, and you want to see the impact before and after that decision.
- Second, intentionally design an intervention with a test group and a control group.
- Third, test multiple things at once.
- Finally, run tests across every single "people decision" you make.

Although Waber notes that "the furthest along any company is right now is single tests," his framework demonstrates how much more is possible on this front.

Our research suggests that employees are surprisingly open to sociometrics to improve performance—90 percent of survey respondents indicated a willingness to do so. Waber's experience echoes this openness; employees get to see their own data, "which is essentially a Fitbit for your career. You can compare yourself to the team average. And not just averages—let's say I'm a salesperson, and I want to be the best salesperson. Well, do I know what the best salespeople in the organization do? And where are there significant differences?" Sociometrics can provide this type of information to employees, so they can adjust their own behavior to improve performance.

Transparency and Permanence of Digital Content

In addition to managing networks, digital platforms also support diverse ways to share and interact with content.[9] Two features of more advanced collaborative platforms are transparency and permanence of content. Email is typically sent to a particular person or group of people for a specific purpose. It presumes that the sender will know what types of information the recipient is interested in. In other collaborative platforms, the contributions are typically observable by others in the organization in real time (transparency) or later (permanence). Thus, potential recipients can search out certain topics or information on the platform, finding the necessary information even if it wasn't initially intended for them.

Transparency can allow others to benefit from information shared in normal interaction. So-called newsfeeds, which are common in platforms like Twitter, Facebook, LinkedIn, and Slack, allow people to monitor all the interactions that take place within a particular group or topic on the platform. Just by scanning these feeds and learning what topics people talk about and are interested in, employees obtain a better idea of the expertise and knowledge of their colleagues and are able to access that knowledge when it is needed at a later time.

Permanence also allows this information to be used by others at a later time. In the BASF example in chapter 11, the teams found that when project teams use digital platforms for collaboration, it is easier for new members to get up to speed quickly with the project team. Although certain members may have left the team, the record of their conversations, decisions, opinions, and feedback are preserved for new members coming in. The new team members can use this information to better build on the previous work, rather than having to re-create it.

Transparency and permanence of content is not sufficient. Content is valuable to the organization only to the extent that it is actually used. For example, we studied the use of an enterprise collaboration platform at a manufacturing company. While the platform appeared to be wildly successful, with new content being contributed every day, the data told a different story about what was really happening. Many people were posting far more content than they were actually consuming. One group of employees, whom we called the "super-promoters," were posting thirty-five times the amount of content they were consuming, in the name of "personal branding."[10] In fact, most of the real collaboration that takes place on the platform occurred in private groups, because the volume of content on the public spaces is overwhelming and often co-opted for other purposes.

Transactive Memory at Discover Financial Services

Both transparency and permanence of content support an organization's *transactive memory*. Transactive memory describes the extent to which people are aware of who knows what in an organization, so that they are able to access that knowledge when needed. The former CEO of Hewlett-Packard notes that if "HP knew what HP knows, we would be three times more productive." Transactive memory is a measure of how well the individuals in your organization know what one another knows, and it has been positively associated with organizational performance.[11] The ability to manage networks and share digital content enabled by advanced collaborative platforms can increase an organization's transactive memory in ways email does not.

Paul Leonardi, of UC Santa Barbara, was studying how employees communicated using the social media platform Jive at Discover Financial Services. Jive provides a newsfeed display of communications happening across the organization. Discover wondered if and how these tools could be useful for communication, and they wanted to figure out how to get employees to use them. Leonardi gave the employees a survey about who knew what and who knew whom in the organization before adopting Jive and then again after using it for about six months. These questions tested the employees' transactive memory, which previous research has shown is important for being able to find necessary knowledge in the organization when needed.

Leonardi found that, from before using Jive to six months after, employees improved their accuracy at identifying who knows what in the division by about 30 percent. They also improved their accuracy at identifying who knows whom by about 88 percent. Yet, when he asked the employees whether

(continued)

> they had learned anything with Jive, they responded that they hadn't learned anything with the system. "No, I didn't learn anything."
>
> So, there is a paradox of advanced collaborative platforms for communication and collaboration. People learn a lot by just becoming aware, by proactively scanning the environment without any idea that anything they were gleaning would be useful in the future. It's a different way to acquire knowledge than we are typically used to through tools like Google search or email. Usually we search when we run into a problem, and we go looking for a solution, so we're looking for something that can fill an immediate need.
>
> These people were not doing that at all. They were slowly becoming aware of what and whom their coworkers knew and filing that meta-knowledge away for later use. So, it didn't register to them that they were actually learning.

Toward Intentional Collaboration

Nevertheless, collaboration simply for the sake of collaboration is not particularly valuable, and unintentional collaboration can often devolve into unhelpful patterns. Some research suggests that digital platforms can lead to similar types of unproductive collaboration if pursued without intentionality. For example, people typically want to interact with other people who are most like them (a characteristic known as *homophily*) and who share common social relationships (a network characteristic known as *balance*). Connecting with like-minded people is enjoyable—but it often reinforces existing biases and degrades effective decision making. When someone connects with people who share similar perspectives and relationships to their own, those new connections typically don't offer new insights or alternative viewpoints that the person couldn't have accessed before. In fact, it creates an "echo chamber," in which individuals become overly confident in their perspectives and decisions, which can be detrimental to individual and organizational performance.

The psychologist Irving Janis noted just how these types of natural collaboration tendencies can go wrong in a phenomenon he called "groupthink." He noted that groups tend to make collectively bad decisions under certain types of conditions, such as homogenous groups that are poorly structured.[12] In fact, these types of groups often make worse decisions than the group members would make individually. Interestingly, these conditions are often the tendencies that humans tend to gravitate toward and that more advanced collaborative platforms can unintentionally reinforce.

In his book *The Wisdom of Crowds*, James Surowecki reverse engineers Janis principles of groupthink to identify conditions under which groups can come together and make stronger decisions than as individuals.[13] These conditions are those in which

- the group has a diversity of opinion;
- group members make decisions independently from one another;
- decentralization enables individuals to draw on local knowledge; and
- there is an appropriate mechanism for aggregating individual opinions to make decisions.

While new collaboration tools can enable more productive collaboration, they will only do so if they are used intentionally, cultivating diversity of opinion, independent decision making, and decentralized communication. It is worth noting here that you need diversity in your organization to have diversity in your collaboration tools. In Surowecki's model, collaboration platforms are just one aspect of the equation for collaboration—aggregation. Without also seeking to develop the other aspects, new platforms can just as easily devolve into groupthink.

Professor Tom Malone, of MIT's Sloan School of Management, refers to this phenomenon as *collective intelligence*.[14] He defines collective intelligence as "groups of individuals acting collectively in ways that seem intelligent."[15] Malone and his group include computers in the mix of collaboration. They ask the key research question, "How can people and computers be connected so that—collectively—they act more intelligently than any person, group, or computer has ever done before?" So, the goal of intentional collaboration is collective intelligence, how people work together through and with technology to make better decisions than any would have been able to make independently.

Performance Benefits Come Second

Some researchers have found, however, that the instrumental benefits of collaboration are often secondary effects that occur *only* after the company used the platform to strengthen the relationships in and the culture of the organization.[16] Their research shows that companies need to first work on cultivating authenticity, pride, attachment, and fun among employees through the platform before they will ever deliver on the type of performance and other collaborative benefits companies are looking for. Digital platforms can be used to humanize a company's leaders, allow employees to be recognized for their accomplishments, cultivate deeper connections between employees and key stakeholders, and allow all members of the company to engage in a certain, appropriate level of frivolity at work. Companies that attempt to derive the performance benefits before developing social capital often end up with an expensive but poorly used platform and underwhelming results. Companies that use platforms first to increase a sense of community among their employees often realize performance benefits as a result of those gains.

In other words, collaboration platforms are still just tools. They will not automatically fix weak relationships or a toxic culture. In fact, they can amplify an organization's culture, regardless of whether it is positive or negative—making a good culture better and a bad culture worse. To reap the most value from these tools, an intentional approach needs to be taken to understand organizational and interpersonal dynamics and cultivate and model the types of communities, relationships, and work desired from these tools. These findings resonate with our own research, which investigated the barriers most organizations faced when collaborating.

We asked respondents an open-text, free-response question about the biggest barriers to collaboration they face at their organization. These responses should probably come as no surprise at this point of the book. The biggest barriers respondents reported are primarily organizational— silos, culture, fragmented departments, time, reluctance to change, and leadership topped the list (figure 13.2). These issues are not likely to be

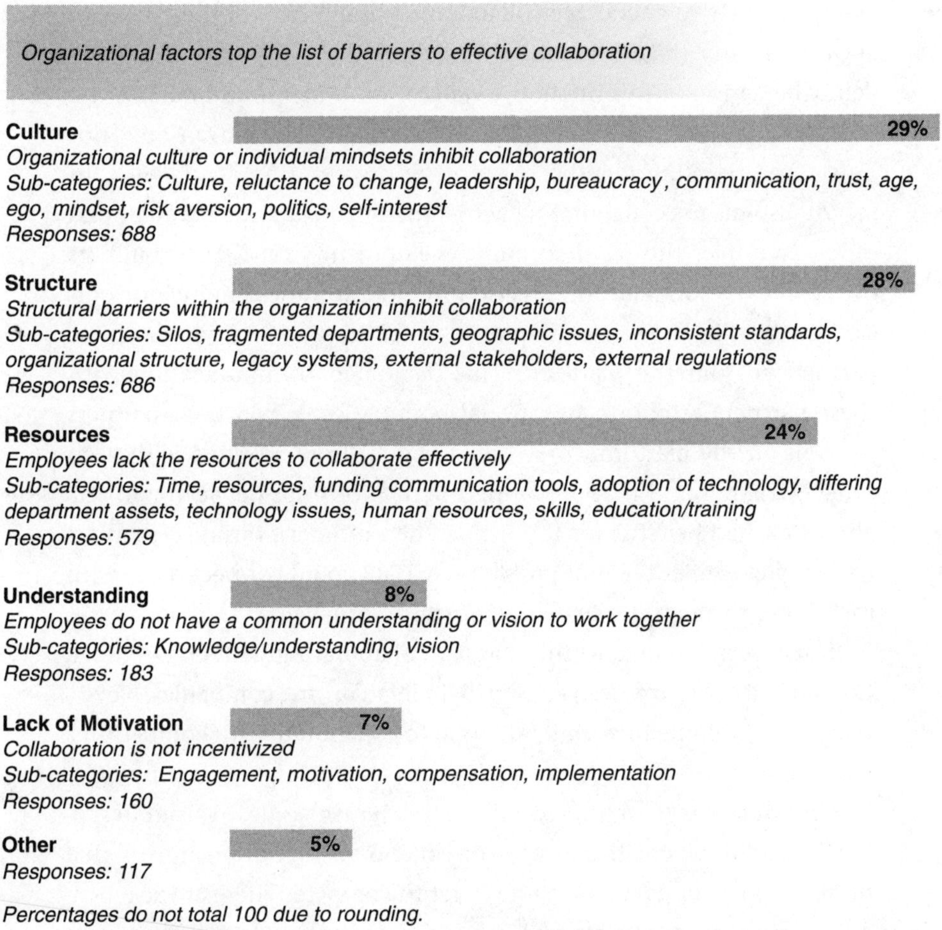

Organizational factors top the list of barriers to effective collaboration

Culture   29%
Organizational culture or individual mindsets inhibit collaboration
Sub-categories: Culture, reluctance to change, leadership, bureaucracy, communication, trust, age,
ego, mindset, risk aversion, politics, self-interest
Responses: 688

Structure   28%
Structural barriers within the organization inhibit collaboration
Sub-categories: Silos, fragmented departments, geographic issues, inconsistent standards,
organizational structure, legacy systems, external stakeholders, external regulations
Responses: 686

Resources   24%
Employees lack the resources to collaborate effectively
Sub-categories: Time, resources, funding communication tools, adoption of technology, differing
department assets, technology issues, human resources, skills, education/training
Responses: 579

Understanding   8%
Employees do not have a common understanding or vision to work together
Sub-categories: Knowledge/understanding, vision
Responses: 183

Lack of Motivation   7%
Collaboration is not incentivized
Sub-categories: Engagement, motivation, compensation, implementation
Responses: 160

Other   5%
Responses: 117

Percentages do not total 100 due to rounding.

Figure 13.2

resolved through the adoption of a collaboration tool alone, but they can often be resolved or alleviated through intentional use of a collaboration platform.

Collaboration beyond the Enterprise

Many digitally maturing companies are also thinking about collaboration across the boundaries of the organization—such as with customers, partners, and even (gasp) competitors. Digitally maturing companies are more likely to encourage collaboration along each of these boundaries than early stage companies. Collaboration today arises from the democratization of information, available both inside and outside the organization. As a result, we have found that collaboration does not only mean working together within the organization across silos. In fact, it also means collaborating across the organizational boundaries.

For example, the research and development firm MITRE built its own collaboration platform precisely because it wanted to include outside partners in the space. Employees were allowed to invite business partners to join the platform. This facilitated communication with these partners across organizational boundaries. When their partners were all on the platform, they could benefit from working with each other. MITRE found that brokering the relationships between partners also often had benefits for the firm. When different business partners were facing similar types of problems, MITRE could connect them with each other through the shared platform, adding value.

It may seem strange to think about collaborating with competitors. Certainly, it is a rarity, even among digitally maturing companies. Nevertheless, some companies find it helpful to collaborate with competitors, particularly when figuring out applications for cutting-edge technologies that no one has yet mastered. The rationale we've heard from executives who talk about these types of collaborative relationships is that no one understands it well enough yet to compete. There are also benefits of smaller companies working with competitors to establish standards that enable them to challenge bigger rivals. Even large companies often don't compete with one another on every front, so partnerships in these noncompetitive areas may be mutually beneficial. For example, in the early days of enterprise social media, one executive reported attending a regular meeting with a group of colleagues in similar functions at different companies. When asked about why he met with these competitors, he replied, "We're all still just trying to figure it out now. We're not ready to compete yet." Just as employees go to conferences to share and learn best practices in other fields, so they collaborate on digital transformation.

The key message here is not that everyone needs to go out and start collaborating with competitors. The message is that companies may need to start thinking about collaborating more broadly, beyond organizational boundaries. The technological infrastructure is now in place to support these types of collaborations, and managers might begin thinking about whether new ways to collaborate would allow them to do work differently.

Cardinal Health: Collaboration "Fuses" Innovation and Culture

Cardinal Health, a global integrated health care services and products company, enhanced its culture by establishing a new innovation center (in 2014) called Fuse. The Fuse team is housed two miles away from corporate headquarters in Dublin, Ohio, and places a premium on cross-functional work.

Cardinal Health established its Fuse Innovation Lab to bring together physicians, patients, pharmacists, and providers. These ecosystem partners work with Cardinal Health innovators to understand issues deeply, craft solutions, and try them out. Brent Stutz, senior vice president of commercial technologies and chief technology officer of Fuse at Cardinal Health, recalls being in the lab recently and seeing one of the Cardinal Health developers in scrubs because he was shadowing clinicians at a local hospital. "It's not just about bringing people in," Stutz notes. "It is about getting them out and observing. We experiment with our customers running week-long innovation and design sessions at their location."

Cross-functional teams including Cardinal Health engineers, creative designers, and scientists work alongside pharmacy customers, health care providers, and other ecosystem participants. Customers and employees regularly propose ideas that are screened and tested using agile one-week sprints. Innovation centers can sometimes be thought of as outposts that are disconnected to the overall company. Cardinal Health's leaders' continued support helps ensure that Fuse initiatives are integrated successfully into the broader organization. "Senior leadership came up with the idea of Fuse, supports it and talks about it," says Stutz. "Without that, I'm not sure we would have the buy-in or the engagement from the rest of the organization or even from our customers."

Stutz stresses that maintaining (rather than just building) a supportive culture is critical. Supporting collaboration can be one of the more demanding challenges. Collaboration needs the right talent. When it comes to hiring for and building a supportive culture, Stutz looks for characteristics such

(continued)

Cardinal Health: Collaboration "Fuses" Innovation and Culture (continued)

as empathy, problem solving ability, curiosity, and adaptability. As he puts it: "It's not always the smartest person we hire, but the person who is going to be the team player and bring a genuine passion and energy for solving big problems."

Building Your Organization's "Digital Tower"

We started this chapter with the story of Catalan castellers. Dozens of male and female participants work together to build human towers up to ten levels high. Intentional collaboration and practice are essential to a casteller team's success. Unlike castellers, organizations facing digital disruption can use technology to build the digital equivalent of a human tower. In a world disrupted by technology, intentional collaboration and practice are also essential to an organization's successful journey toward digital maturity. Hierarchical structures with solid lines on org charts will continue to diminish, and organizations will look more and more like peer-to-peer networks. Power through simple authority will rarely exist. Instead, influence and persuasion become key to building support and getting things done.

Collaboration platforms can help, both within the organization and beyond its boundaries, by helping individuals connect more purposefully and by making content more broadly accessible. But leaders at all levels must use these tools intentionally to help the organization increase its awareness of "who knows what" and raise the organization's collective intelligence quotient. Like a human tower, your organization's "digital tower" will take strength, balance, courage, and common sense.

Takeaways for Chapter 13

What We Know	What You Can Do about It

• Engaging in intentional collaboration is a key characteristic of digitally maturing organizations.
• Digital platforms change collaboration by providing two key capabilities:
  ◦ managing networks
  ◦ sharing digital content
• Benefits of collaboration include:
  ◦ transactive memory
  ◦ collective intelligence (better decision making)
	• Select a collaboration platform that fits your needs. Include mobile capabilities as a selection criterion.
	• Create groups for general communication and collaboration, as well as specific groups built around work or special topics.
	• Pilot the use of the collaboration platform. To guard against biases and reap the benefit of collective intelligence, create these conditions in groups:
	  ◦ Cultivate diversity of opinion.
	  ◦ Monitor for groupthink and allow for independent decision making.
	  ◦ Decentralize communication.
	  ◦ Provide an appropriate mechanism for expressing and aggregating individual opinions to make decisions.
	• Leaders should participate in using and contributing to the platform.
	• Select "collaboration champions" who can provide momentum and assistance to help others use the collaboration platform effectively.
	• Consider setting up a collaborative center of excellence to increase momentum, adoption, and focus.
	• Move conversations from email to your collaboration platform.

Fail fast. The mantra of innovators and entrepreneurs. Articles, blogs, and books tout the importance of failing fast, failing early, and failing often.[1] The Silicon Valley zeitgeist is all about rapid prototyping, releasing minimal viable products, quickly finding flaws and correcting them, and celebrating failures as the necessary precursors to success. In this world, we love a failure that gives rise to a pivot, and which, in turn, allows an organization to recast itself as something entirely different from what it had been nanoseconds ago.[2] The question is, How can this concept be successfully applied by would-be digital innovators in legacy organizations, particularly in organizations designed to reduce or eliminate failure? That's the focus for this chapter.

Let's start with a bit of context from our research. It should come as no surprise that digitally maturing companies innovate better than less mature companies. We asked respondents how their companies are innovating relative to their competitors. While only about 20 percent of respondents from the earliest stage companies described their companies as innovative, the number approaches 90 percent in digitally maturing companies. This innovation does not happen accidentally. Respondents reported that digitally maturing companies are far more likely to invest in innovation: while 87 percent of respondents from digitally maturing companies said they invest in innovation, only 38 percent of respondents said early stage companies do so. Being innovative, however, is not simply about doing innovative things. Instead, it is about cultivating an organizational environment that is conducive to innovation. It is

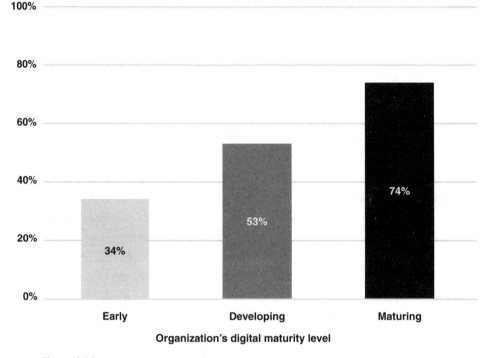

My organization encourages new ideas to be shared and tested at all levels of the organization (Respondents who answered "strongly agree" or "agree")

Figure 14.1

about being open to new ideas, wherever they may be found. Respondents from digitally maturing companies were more than twice as likely as those from early stage companies to encourage new ideas to be shared and tested at all levels of the organization (figure 14.1).

Perhaps more important, being innovative is also about a willingness to act on those ideas. When employees were asked whether their managers encouraged them to innovate with technology, the responses charted a similar pattern, ranging from around 20 percent for people at the earliest stage companies to over 80 percent for digitally maturing companies. Leaders of most organizations would agree that, in theory, innovation is critical to business success in a rapidly changing environment. In practice, most organizations that were not born in the digital age struggle mightily with innovation for two reasons:

1. The cultures of most legacy organizations have evolved to reduce or eliminate the variations necessary for the experimentation that, in turn, leads to innovation.
2. Leaders find it challenging to innovate while also keeping the company's core businesses running efficiently and effectively.

Legacy Companies Are Built to Eliminate Experimentation

We asked respondents about the biggest challenge affecting their organizations' ability to compete more effectively in a digital environment (figure 14.2). The top answer, by a wide margin, is experimentation and getting people to take risks. Interestingly, this barrier is the highest for

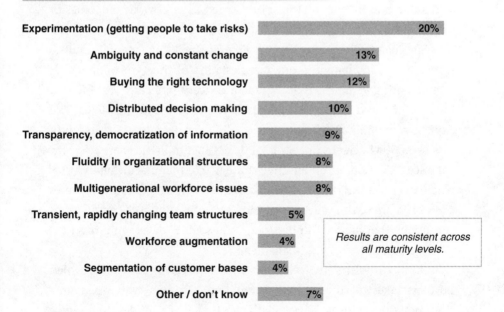

Figure 14.2

each maturity level and relatively equal across all maturity levels. Even companies that are innovating better than their competitors still say that experimenting and getting people to take risks is the single biggest challenge they face.

What makes experimentation so hard in most organizations? Quite simply, it flies in the face of what most companies have been built to do over the past fifty years—optimize efficiencies and minimize operational variances. We've seen traditional companies struggle with the need for experimentation, because they are driven by a fear of failure. It's fascinating how some young digital companies experience failures "every single day" in their efforts "to achieve their purpose, and they're comfortable with it," says Maile Carnegie, group executive, digital banking, at ANZ. That comfort "starts with the audacity of their mission," whereas many established companies have a "fear of failure baked right into" their culture.

She continues: "In a company like Google, its purpose is to literally change the world. The company holds itself to a lofty and unachievable mission. If you compare that to a lot of legacy companies, they have achievable and incremental missions. One of the obvious outcomes of that is if you're trying to strive for something that is incremental, you'll hit your goal, but you are also going to get small, incremental results." As we discuss in chapter 1, one manifestation of this type of approach to eliminating variances is Six Sigma, a concept pioneered at Motorola in the 1980s and championed by Jack Welch at GE in the 1990s. The explicit goal of Six Sigma is to reduce all variance and defects in a process to 0.00034 percent or about 3.4 defects per million. These efforts at variance reduction are an effective strategy in a manufacturing industry where the organizational goals and conditions are stable. Yet, companies can only achieve this level of precision by optimizing to a certain set of conditions that are well understood. Thus, GE is developing the Fast-Works program we discuss in chapter 1.

In more turbulent environments, where the conditions under which business takes place are rapidly changing, efforts at variance reduction may not be ideal. They may have the benefit of identifying new business processes or opportunities that are far more effective or efficient, but a possible unintended consequence is that the organization may not be able to realize these benefits because it is optimized around the old conditions. In many ways, a culture of experimentation is about intentionally creating variance in existing processes to see if there are better ways of doing things.

Certainly, Google is quite different from most companies. Their offerings lend themselves to this type of experimentation, and they have abundant resources to devote to these efforts. Yet, the core lesson still rings true. Running experiments is essential if your company wants to understand what is possible.

Experimentation at Google

Richard Gingras, vice president of news at Google, is no stranger to digital disruption and transformation. He was a pioneer of digital journalism, having founded Salon.com, and worked in the news industry throughout its digital disruption. He has been instrumental in Google's Accelerated Mobile Pages (AMP) project, helping make the internet more usable in a mobile age.

A key approach to successfully navigating digital disruption is experimentation. He says, "Deep within our approach to product development is rapid change and experimentation. There isn't a team that I'm working with which isn't right now running anywhere from a handful to dozens of experiments with the user base. You're just constantly iterating, iterating, iterating, and running experiments."

Experimentation is challenging in most traditional companies because they don't have the necessary mindset for experimentation. Gingras continues, "It's hard because the long history of corporate cultures was not necessarily set up [to] stimulate innovation or experimentation. They're typically set up to avoid risk. As I pointed out, if you look at the history of corporate culture, well, the whole approach is about how do I improve consistency, grow margins, and eliminate risk."

Even when traditional companies do innovate, they usually think too small, benchmarking themselves against competitors, who are also too risk averse. He says, "You don't innovate based on what you think your competitors are doing; innovate based on where you think you can be five years from now. You're looking for the 10X improvements in what you're trying to do

(continued)

Experimentation at Google (continued)

because if you simply site yourself based on your competitors, then you tend to do incremental stuff. You're not really thinking about where you might be."

The problem is that traditional companies think of experimentation in terms of success or failure. Instead, the outcomes of an experiment should be measured in what was learned, not whether it succeeded. Even a successful experiment isn't helpful if the company doesn't learn something from it. Gingras concludes, "It's really not that important if the experiment succeeds or fails, it's what they learn from it—good, bad, or indifferent, it is intelligence that they can lay claim to. Maybe it didn't work out the way we thought it would but we learned X, Y, and Z, and we're not embarrassed by the fact that our initial assumptions were wrong. There are no failures. We tried something and we learned something."

Test Fast, Test Small, Test Enough

A primary reason that many organizations struggle with experimentation is that they have conditioned themselves to believe that failure is anathema. No wonder most organizational employees are nervous about experimenting if the company's tolerance for failure is the Six Sigma threshold. You cannot hope to successfully experiment and still only fail 3.4 times out of every 1 million attempts. In the digital age, how companies deal with setbacks may determine their ability to survive. Because new challenges are becoming the norm, and so much is unknown and untested, failure is inevitable. So a critical factor in helping organizations become better experimenters is helping them become better at testing ideas, learning from these tests, and scaling quickly when the tests reveal productive insights. While they need to develop an environment of productive failure, the fail fast mantra of innovators and entrepreneurs may be viewed skeptically by organizations conditioned to detect and eliminate failure. It may be easier to focus the organization on adopting a "test and learn" mindset, rather than a fail fast one.[3]

A good way to "test fast" is to put a fixed short-term timeline on experiments. Experimental organizations often perform short "sprints" (e.g., six-to-eight-week initiatives) during which they attempt to change one aspect of the organization. At the end of the sprint, the experiment is concluded, and the success or failure of the experiment is determined. This established time frame provides managers the easy decision point to stop or refocus the experiment, instead of allowing faltering projects to be drawn out for long periods—a common occurrence in many organizations. An important addendum to the test fast mantra is also to "test small." A company wouldn't want a multibillion-dollar IT implementation project to fail just to learn some important lessons. So, the company must ensure that it creates acceptable parameters for experimentation and learning to take place. Be sure you set up small experiments that limit the damage from failures, allowing you to learn and move on.

Last, "test enough." Companies need to manage risk as a portfolio and keep failure within a certain tolerance level. Is the right failure rate 10 percent or 90 percent? We've heard both from different leaders in different organizations, but make sure to find your organization's Goldilocks Zone for testing.[4] Yet, managers should also remember that if they aren't failing enough, they may not be bold enough. For example, one division of the US Department of Agriculture has an established risk-tolerance threshold for digital projects. If they aren't failing enough, they increase the ambitiousness of the projects to fail more often. Not many think of the government as a place to look for innovative practices, but we think that many companies can benefit from an environment of testing enough.

Learn Faster

Though the word "failure" still has a negative connotation, dialogue around it is starting to shift. Even so, all the talk about organizations needing to fail fast emphasizes the speed aspect and deemphasizes the *learning* aspect. The idea is not simply to fail quickly and move to the next idea. Insights must be gathered from the failure to make it valuable. Just knowing that "A didn't work, so let's try B" is not enough. *Understanding why* A didn't work is where the insights are found and where the learning happens for the organization.

As scientists will tell you, they conduct experiments to test hypotheses and ideas. By proving or disproving their hypotheses, they gain knowledge. In a similar way, organizations should approach experimentation with a goal to learn. From this perspective, testing (along with failure) becomes valuable as input or insight into what didn't work and what could have been done better. The key is to not get stuck in these setbacks and instead learn from them and move on. As Thomas Edison said, "I have not failed. I've just found 10,000 ways that won't work." Indeed, our data suggest that digitally maturing organizations engage in this type of learning from experiments more so than less mature companies do (figure 14.3). While only 21 percent of respondents from early stage companies agreed that their leaders share results from failed experiments in constructive ways that increase organizational learning, 56 percent of respondents from digitally maturing companies reported the same.

The implication of this approach is that you should also learn from projects that succeed.[5] If you don't understand why a particular experiment succeeded, then you don't know whether and how to export lessons from the successful experiment to other projects and settings. Knowing why you succeed may be just as important as the success itself. Success is not the short-term goal, but learning is. Just as learning is essential to the digital talent mindset, it should also be a key part of the organizational mindset.

The process of learning at the organizational level can take many different forms. For some, it can involve formal after-action reports or sessions that debrief the results of an experiment to extract lessons learned. For others, it may involve all-hands meetings where project teams present ongoing results and gain feedback from members of other teams. Learning could simply involve employees not on the team examining the team's electronic data on collaboration platforms for ideas or information generated as a part of the project. For still others, it may involve splitting the team up and incorporating its members into new teams to combine employee knowledge in new ways. In short, organizations can learn from their successes and failures in several different ways, but we bet that if you can't clearly identify the ways in

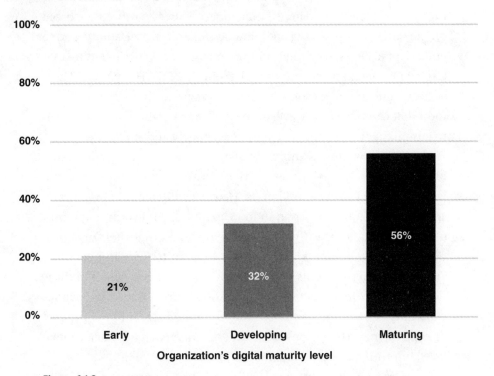

Leaders in my organization share results from failed experiments in constructive ways that increase organizational learning
(Respondents who answered "strongly agree" or "agree")

Figure 14.3

which learning is intentionally happening in your organization, it's not. The key is to clearly identify and intentionally engage in processes that help the organization learn from its experience.

Iterate, Iterate, Iterate

As scientists remind us, lessons learned should not occur in a vacuum. Instead, learning during one round of experimentation should inform the types of experiments companies engage in during the next round of innovation. The experiments and failures should not be scattershot unrelated projects, but targeted toward a specific overarching goal. This discipline becomes even more important in the digital age because of the opportunity for machine learning and AI, an area in which Capital One is also a pioneer.[6] For example, Capital One is using machine learning to customize content in real time for every user on its website, based on how they behave during an online session.[7] Zachary Hanif helped to build Capital One's machine-learning center of excellence to ensure that Capital One could deploy machine learning at scale. Hanif says, "There is great value in having some kind of rich reproducibility tracking platform that can trace nearly every aspect of each iteration of new models to ensure that results can be deployed accurately and efficiently."[8]

Just as innovation is more about the environment of cultivating ideas at all levels of the organization, so the spirit of feedback and learning should permeate all levels. Here, the difference between early and maturing companies is even more extreme. While only 34 percent of respondents from early stage companies said that their organizations encourage feedback and iteration to learn to work in new ways, 76 percent of respondents from digitally maturing companies said the same. Iteration involves taking the lessons learned from the previous experimentation efforts and building that knowledge into the next set of experiments. The lessons from experiments may also be at a meta-level, informing the experimentation process.

Some respondents disagreed with each other about the degree to which this feedback and iteration is happening in organizations, with a significantly higher percentage of c-suite executives than employees lower down in the organization reporting that their companies use feedback and iteration to learn how to work in new ways (figure 14.4). Between 54 percent and 79 percent of c-suite executives reported that they are using feedback and iteration to learn how to work in new ways (with the CEO or president at the top of this range), but only 38 –51 percent of lower-level employees reported that their company is doing so. These results may mean that the iteration is happening at a higher level than the typical employee is aware, or perhaps more likely, that the efforts aren't having as significant an impact on work as the executives believe.

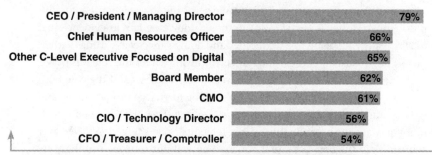

Figure 14.4

Scale Fast

Another danger with the fail fast approach is that companies are satisfied with simply running experiments and think this is enough to embrace risk. Yet digital initiatives aren't risky if they never promise to touch your core business. And, considerable research suggests that innovation has little real effect on business unless companies fundamentally transform business models.[9] Experimentation just for the sake of experimentation isn't particularly valuable.

John Hagel, of Deloitte's Center for the Edge, notes that this disconnect is common in many organizations. Companies may engage in innovation efforts, but often these innovations never affect business.

> Every large company has an innovation lab here in Silicon Valley. I joke that I go up and down the street and every large company in the world has an outpost here. And they'll always point to those innovation labs as evidence that they're really embracing this technology and doing very creative things with it. But, so far, it's just an outpost and has very little impact on the core business.
>
> There is an unstated agreement between these outposts and the core business. And the unstated agreement is we're going to give you some money and we're going to give you some space and resources and people. You can do whatever you want in those sandboxes. Just remember one thing; don't ever, ever come back to the core because if you do, you will not survive. And these are very reasonable, rational people, and they understand the rules of the game.

Our research suggests that a key difference that sets digitally maturing companies apart is that they take those experiments to the next level. We found that digitally maturing companies are over three times more likely than early companies to take experiments and then roll them out across the enterprise, instead of limiting their initiatives to small stand-alone experiments (figure 14.5). Digitally maturing organizations are also nearly twice as likely as digitally developing companies to report scaling initiatives.

Indeed, the key differentiator of innovation at digitally maturing companies may not be their ability to experiment well and learn from their experiments. The more significant differentiator may be their willingness to take the lessons learned from both successful and failed experiments and scale those experiments across the organization to drive business model transformation.

Changing the Wings Midflight

The second reason that companies struggle with innovation is that they can't afford to drop everything and shift their focus to learning and experimenting with new technology. They must innovate in a way that allows their core business to function while also influencing that core business. John Halamka, BIDMC CIO, equates it with trying to fix a plane in midflight. He says, "It is just really hard to be a digital

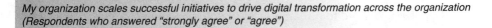

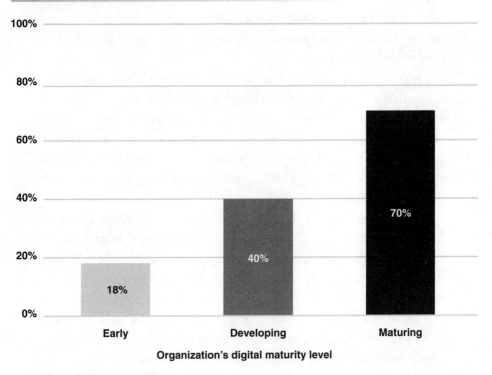

My organization scales successful initiatives to drive digital transformation across the organization (Respondents who answered "strongly agree" or "agree")

Figure 14.5

leader in the current environment when you are being asked to change the wings on a 747 while it's flying. Total security, total reliability and stability, with complete innovation at the same time. I wonder what people will want to take this role? Or maybe the role has to be recast and . . . the tasks [divided] across multiple individuals so they can deal with the pace of change and the stress."

Digitally maturing companies have found ways to be innovative amid the need to maintain business operations. Some of this comes from exactly how they innovate digitally. As companies begin to experiment more, this increase puts additional pressure on the need to balance experimentation with effective exploitation. In a foundational paper on organizational learning, James March notes the importance of balancing exploration and exploitation in organizational learning.[10] Organizations need to find new ways of doing business through exploration and experimentation while also maintaining a viable business and exploiting established competencies. O'Reilly and Tushman referred to companies that successfully accomplish this balancing act as "ambidextrous organizations."[11]

Cisco's James Macaulay echoes this need, noting, "One of the key challenges that any large company faces—any large, successful company—when it comes to digital transformation is maintaining your existing business while expanding into new businesses, in some cases where maybe there's some friction between those two. That's something that all large companies must manage as they innovate and try to disrupt themselves in the most positive sense of the word."

Our survey data support Macaulay's view. We found that digitally maturing companies were not necessarily more likely to experiment than early stage companies, but they were more likely to report balancing the need to explore new competencies with the need to exploit existing capabilities. We asked companies what the purpose of their digital initiatives are. Early stage companies were over three times more likely to report that the purpose of their initiatives is either entirely or mostly about exploiting existing organizational competencies.

Volvo Cars Balances Competing Concerns for Digital Innovation

Working on a strategy for connected cars, Volvo's executive team outlined a vision that would "give life" to cars beyond the time of production.[12] New technology would enhance end-user experience and open up new revenue streams. By disconnecting from traditional automotive cycle plans, the car connectivity could increase the pace of change. It would allow the firm to engage with external innovation ecosystems and sync with developments in consumer electronics. Leveraging connectivity by exposing the car to external developers through open APIs could also generate a new level of functional diversity in the automotive industry. Such APIs could, for example, make the mobile phone a dynamic front-end to the car.

Efforts to introduce a connected car required several different shifts in innovation, governance, and partner relationship management. The shift to more digital operations required executives to think of their product differently—they were not simply upgrading the cars themselves but creating a platform on which others could innovate. They required new types of innovation processes to make the platform happen within the organization and to encourage others outside the company to develop for the platform. These new types of outside partnerships also required more flexible ways of contracting than what Volvo was using with traditional partners, usually based on monetary transactions. Since Volvo was not paying Spotify or Pandora to develop apps for the platform, for example, the new contractual relationship needed to encourage the types of value cocreation both companies sought.

Furthermore, this digital strategy faced the challenge of keeping existing business processes running while also trying to innovate. For instance, when Volvo executives sought to develop the initiative around connected cars, the process led to major changes not only to the technology but also to multiple aspects of the business that they had not previously considered. They needed to balance competing concerns between innovative digital practices and established business processes across innovation capabilities, process versus product focus, internal versus external innovation, and flexibility versus control. Volvo's digital initiative succeeded only by effectively balancing these competing concerns.

Funding Innovation

Larger companies also need to realize that simply throwing more money at the problem is not the answer. In fact, lack of money may be an advantage on this front. John Halamka of BIDMC says, "It helps if you're underfunded. It forces you to be scrappy. That's a big part of the culture." Although he works at a $5 billion organization, the IT budget is 1.9 percent of the organization, forcing him "to be edgy and innovative." Kimberly Lau of the *Atlantic* describes a common situation at midsized firms: "You have to be focused on where you make your bets. We have to choose carefully, because we've got limited resources." It's imperative, then, to work "fast and move on quickly, because there's always somewhere else to put those resources."

A key challenge to playing the long game is finding resources to move initiatives forward while tending to the existing business. Many companies are finding investment capital through shrewdness and discipline. At Marriott International, for example, George Corbin, a former senior vice president of digital, says that some of the company's most important innovations are its funding models. "Finding ways that we can make a growth opportunity become self-financing is critical," he says. "If I can do that, the digital opportunity can stand on its own two feet and scale sustainably."

Even though companies can't just throw money at the challenge of digital transformation, it doesn't mean that funding isn't a key challenge. We asked survey respondents whether funding digital initiatives was a significant challenge that affects their organization's digital efforts. Sixty-five percent of respondents agreed that it was a significant challenge, 18 percent disagreed, and 17 percent neither agreed nor disagreed. These numbers were relatively consistent across maturity groups. So, how do companies solve these funding issues? Some companies set up venture capital funds within the organization to fund these internal innovation efforts. Others reinvest savings gained from a previous round of digital initiatives to spur new innovation. Regardless of the approach, leaders need to think through how to fund digital innovation, even if they have to get creative with how to do so.

Double-Loop Innovation

In 1977, Harvard professor Chris Argyris introduced the concept of "double-loop learning," which describes the need for organizations to not only learn but also develop new ways of learning. We might extend that concept to cover double-loop innovation. Competing successfully in a digital environment might require creating new ways of innovating, which can touch all aspects of the organization. All this suggests the need for a purposeful and disciplined approach to digital innovation. Years ago, and long before the words "digital disruption" entered our lexicon, management guru Peter F. Drucker authored an article on innovation for *Harvard Business Review*. While the context has changed dramatically, the advice (like much of Drucker's advice) is as sound as ever: "To be effective, an innovation has to be simple, and it has to be focused. It should do only one thing; otherwise it confuses people. . . . Effective innovations start small. They are not grandiose. . . . [W]hen all is said and done, what innovation requires is hard, focused, purposeful work. If diligence, persistence, and commitment are lacking, talent, ingenuity, and knowledge are of no avail."[13]

Takeaways for Chapter 14

What We Know	What You Can Do about It
• Innovation is critical to survival in the rapidly changing digital age. • Risk taking, experimentation, and failure are all necessary elements for achieving innovation. • Most organizations born in the twentieth century are optimized for efficiency and productivity—and designed to eliminate variation, reduce experimentation, and minimize risk. • Organizations need to learn how to conduct pilots in which failure is a possible outcome. The goal is not failure per se but rapid learning and adaptation.	• Assess your organization's risk appetite and identify where there are barriers. • Determine which risk and security policies are nonnegotiable and which have flexibility of modulation. • Make your customers aware of selected risk and security policies. • Before engaging in risk experimentation, determine and communicate guiding principles of acceptable and unacceptable risk taking. • Put a fixed short-term timeline on most experiments, such as conducting short "sprints," or six-to-eight-week initiatives, that are assessed at the end of each cycle. • Release the minimum viable product to a small group of trusted customers and/or stakeholders for feedback. • Identify the learning feedback loop process and make it intentional. It is critical to avoid ad hoc iterations that don't build in learning. • Feed learnings and feedback into subsequent versions and continue the iterative cycle. • Communicate approach with selected clients and pull them into selected iterations—identify and communicate guiding principles and approach (e.g., unacceptable and acceptable failures, learning loops, built into subsequent versions).

15 Moving Forward: A Practical Guide

By this point of the book, we hope we have convinced you that it is time for your organization to take the first steps toward digital maturity. The goal of this chapter, consequently, is to provide you with practical and pragmatic guidance for moving forward. We draw on many of the lessons in this section, such as experimentation, becoming intentionally collaborative, and iteration to show how organizations can take these lessons and put them into practice.

Three-Step Process for Increasing Digital Maturity

The process for making practical progress toward digital maturity and ultimately becoming a digital organization involves three distinct steps. Your goal is to reimagine how work gets done at your company, how it aligns with the future of work, and how it sets the stage for cultivating a digitally maturing organization.

1. **Assess.** To understand where your organization needs to go, you first need to understand where your organization stands with respect to digital maturity. While we typically speak throughout this book about digital maturity as an organization-level characteristic, in reality, digital maturity is unevenly distributed throughout your organization. Some divisions, teams, or processes are more or less digitally mature than others. Indeed, we mentioned in chapter 3 that when we present findings from our research to an executive group from one company, members of the group often have different perceptions

about how digitally mature their company is, mainly because of the differences in where they sit in the organization.

In much the same way as the proverbial blind men describing an elephant as a snake, a tree trunk, a wall, or a rope depending on where they touch it, your organization's digital maturity may appear very different, depending on where you experience it. Part of assessing your organization's digital maturity is also about understanding how and where it is distributed throughout your organization. In this chapter, we provide you with tools for assessing how digital maturity is distributed throughout your organization.

2. **Enable.** In this step, you determine how digitally mature your company needs to become today. Organizations can employ two different strategies for identifying the areas ripe for transformation. One strategy is for organizations to identify digital strengths to continue to use toward an in-progress transformation. This approach is probably helpful for more advanced organizations that have already mastered many aspects of digital maturity and may seek to extend advantages they already possess. Another strategy, and one that more traditional legacy companies may find more helpful, is the reverse formulation. The question is not how your company can become more mature, but what are the steps you can take to become less digitally immature? Your company's overall digital maturity may be defined more by the least mature areas than by its average level of maturity because the bottlenecks may hold the entire organization back.

Regardless of which strategy you pursue, the decision on what areas to begin transforming should be subject to a cost-benefit analysis: What will be the financial and organizational cost of changing this aspect of the organization, and what will be the benefits of the transformation if successful? You will want to pursue the aspects of the organization that will yield the best return on your investment of time, energy, and resources.

3. **Mature.** In this step, you determine how your company improves on the target area and practically moves toward digital maturity. In this

step, it is important to retain lessons learned earlier in this section of the book. You won't want to adopt waterfall methods that attempt to change the entire organization at once through meticulous planning and careful enterprise-wide execution. Instead, develop short sprints characteristic of the agile methodology discussed in chapter 12 to make the minimum practical steps toward digital maturity as identified in step 2 above. This process will also likely require the types of experimentation and testing that we outline in chapter 14.

Moving to the Next Stage of Maturity

Although our data have shown that our three-stage model is the best way to describe where companies are in the process of digitally maturing, when practically assessing the organization, we include four possible stages of maturity. Why? Throughout this book, we note that the qualifications for each standard keep advancing along with technology. While a three-stage model may be helpful for understanding where organizations currently *are* along the path to digital maturity, a four-stage model is more helpful for considering where companies should be *going*. Although certain aspects of the organization may be at the third (maturing) stage, the fourth stage of maturity represents the continual opportunity to move beyond where you are to another level of digital maturity. It is possible that some pockets of your organization are, in fact, working to move beyond now. Indeed, throughout this part of the book, we see that digitally maturing companies are also those that invest the most in becoming more collaborative, innovative, and risk tolerant. They would not be making these investments if they thought they had somehow "arrived" at digital maturity.

In other words, the standards for maturity keep changing as technology keeps evolving. Even if an organization has achieved digital maturity today, the changes coming tomorrow will undoubtedly require further change. Thus, the first three stages reflect the three stages of our maturity model, while the fourth represents where companies may be headed in the future.

1. **Exploring digital efforts** (early stage): The organization uses traditional technologies to automate existing organizational capabilities. It may be dabbling with digital, but that has so far resulted in minimal changes to the company.

2. **Doing digital initiatives** (developing stage): The organization is increasingly leveraging digital technologies but is still largely focused around the same business, operating, and customer models existing in the organization today. If the company is maturing, it is largely incidental, incomplete, or isolated. It is focused on supporting digital technologies rather than becoming more digital as an organization. This is an important step in digital maturity but insufficient as a destination.

3. **Becoming digitally mature** (maturing stage): How work gets done in the organization, as well as its interactions with customers, partners, and suppliers, is becoming much more intentional and networked as well as less siloed. Your organization is deliberate in creating more advanced changes to current business, operating, and customer models. Key ways in which the company works, operates, and behaves are changing and shaping a much more mature digital organization.

4. **Being a digital organization** (aspirational goals): Business, operating, and customer models are optimized for constantly changing digital environments and ecosystems. This is profoundly different from traditional business operation and customer models; and digital is at the core of how the firm is organized and how it operates and behaves. Being digital is part of the organization's DNA and not an alternative approach of acting or being.

Assess Your Organization's Digital Maturity Using Digital DNA Traits

The broadest definition of an organization's culture can be gleaned from the way it organizes, operates, and behaves—often referred to as its organizational DNA. Just as your personal DNA defines your traits as an individual, an organization's DNA makes it unique from other

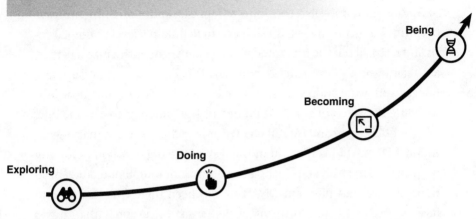

Digital DNA maturity spectrum

Figure 15.1

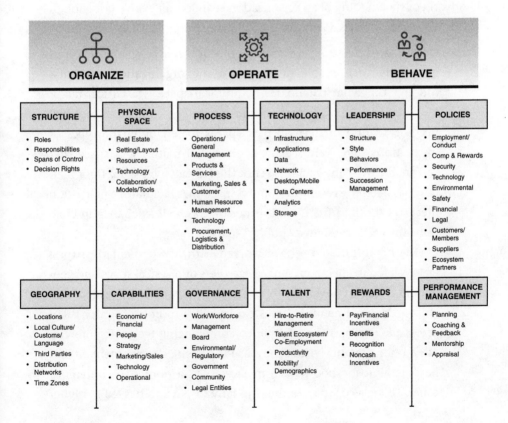

Figure 15.2

companies. How your organization manifests each of these DNA traits is what makes it what it is today.

An organization's DNA continues to replicate and evolve in some ways over time but generally will do all it can to maintain a level of homeostasis and resist all but the most incremental change. As with human beings, an organization's DNA can be strong. This is one reason that so many merger and acquisition (M&A) activities are incredibly difficult or lead to failure as two companies with different DNA attempt to become one. Without deliberate and careful change approaches, some of that organizational DNA will likely stand in the way of the organization becoming more digitally mature, especially in traditional or long-tenured organizations. It is easy to pinpoint DNA that has been there from early on in its existence and is just as strong or perhaps stronger today than it was generations ago. These traits can be positions of strength in this digital world or the undoing of an organization that doesn't evolve fast enough with changing times. Consider the following examples of DNA traits:

- **Organize**—Refers to the structure, physical space, capabilities, and geography that enable or constrain the organization's activity. Indeed, somewhat paradoxically, physical space is a critical factor of digital maturity. Throughout our interviews, companies reported needing to find or develop new spaces for enabling the types of interactions they were seeking with digital maturity. A space that can enable these interactions is not the typical open-space floor plan that has been the recent rage, but a space that facilitates small group meetings (both virtually and in person), as well as chance interactions through redesigned "public" spaces.

   One organization we have worked with has particularly strong DNA related to "being present," regardless of physical space and geographic location—even halfway around the world. Its leaders and employees use a combination of conference rooms (scores of them across the organization) and video conferencing technology to create an "in present" virtual/physically present environment. It has become almost taboo to dial in faceless as a nonvideo conference caller. This is a core part of the company's DNA and has led to higher levels of innovation, greater collaboration, and higher employee engagement scores.

- **Operate**—Refers to the process, technology, talent, and governance that defines the organization. Groups that are spearheading digital maturity initiatives must be given a certain amount of latitude to operate differently than the traditional organization. At John Hancock, the digital teams operate somewhat independently, to free them from the traditional bureaucracy elsewhere in the organization. At BASF, teams banned email in order to learn to use more advanced collaborative tools to work together. These groups need to figure out new ways to operate, and they won't be able to do so if they also need to operate in the old ways.

- **Behave**—Refers to the policies, rewards, leadership, and performance-management mechanisms in place at the organization. Here is really where the proverbial rubber meets the road. Companies often talk a good game about digital maturity, but unless it becomes a critical part of your performance-management and incentive structures, it is unlikely to ever happen. In pushing its own efforts toward digital maturity, Walmart has integrated digital performance metrics into every executive's performance review. People respond to incentives, and you need to be sure your incentives are designed and aligned to support your digital maturity efforts. People won't prioritize it if it is not a part of their jobs.

Each of these levers may help or hinder your organization's digital maturity. Company leaders should think through how different aspects of the organization's DNA contributes to or detracts from the digital maturity of the overall organization needed today and tomorrow. Key questions to ask:

- Which areas of your organization are naturally moving toward digital maturity?
- Where is there the most resistance to maturing digitally?
- How can you build on pockets of success and leverage these across the organization?

In this book, we address some of the most critical digital traits in depth—such as continual innovation, intentional collaboration, iteration, shifts in decision rights, flattening hierarchies, continual disruption, among others. These digital traits make up digital DNA. We suggest that twenty-three traits define digitally mature organizations. This isn't focused on digital technologies but rather the digital DNA an organization needs to operate effectively in this continuously changing future. Digital DNA carries the underlying instructions, development, functioning, and replication for "being" digital.

Your organizational DNA may or may not have these digital DNA traits infused into its structure, governance, capabilities, leadership actions, talent processes, and policies. If you do have these traits, they may or may not be at the level of maturity needed by your organization to thrive in this digital world. It's important for all companies to determine whether they have these traits, and if so, how mature their digital DNA is today. Here, we bring these traits into a concrete inventory by which the organization can begin to assess its own digital DNA maturity.

We suggest two options for assessing your organization's digital maturity, which may be conducted independently or, ideally, together, in either order:

1. **Conduct a survey of employees,** asking them to assess the maturity of your organization's digital DNA. For each of these twenty-three digital DNA traits, leaders and employees rate the company's digital maturity on the 1–4 scale (with "exploring digital" as a 1, and "being digital" a 4). Our data show that middle management and staff-level employees often have a different and less optimistic assessment of their organization's digital maturity. We also often find that different generations of employees, experienced hires, and those in different geographies, business units, and functions can each have different views of the organization's digital maturity. Surveying all (or a representative set of) employees and conducting working sessions with middle management can identify key enablers and uncover barriers to digital maturity that executives aren't aware exist. Instead of

just relying on executives' assessments of an organization's digital maturity, this approach gets all employees' input and paints a more complete picture of the company.

2. **Conduct interviews with executive leaders** to discuss and identify their perspectives of the various levels of digital maturity existing and needed in the organization. Part of these discussions should include understanding your company's digital ambitions. As we mention in chapter 4, this requires *seeing* differently, *thinking* differently, and *doing* differently as it pertains to *being* digital. These leadership discussions will generate information to help obtain a composite view of the organization's current and aspirational digital maturity.

Together these two approaches are more likely to produce the most comprehensive, rich, and nuanced picture of digital maturity because they incorporate feedback from all levels of the organization as well as the digital ambitions of leadership.

Recombinant DNA as a Metaphor for Digital Transformation

When combined with the concept of recombinant DNA, the DNA metaphor is a helpful way of thinking about how companies mature digitally. In gene splicing, one organism's DNA is cut apart and another's genetic material is spliced into that space. The modified DNA is then replicated and reinserted back into the host. The result is recombinant (that is, new or modified) DNA, which then begins to replicate across the organism. The organism may or may not exhibit different characteristics as a result of these changes, but it does include features of the host organism modified by the new traits of the foreign DNA. A common use of recombinant DNA is to develop crops that are resistant to insects or certain types of insecticides. This process also involves a certain amount of trial and error. If you splice in too much DNA, it can overwhelm the host traits. If you don't splice in enough, it won't exhibit enough or the extent of the traits you want. The process is iterative, and it's getting considerably more advanced.

In many ways, gene splicing and recombinant DNA is an excellent metaphor for how organizations can digitally transform. Leaders identify some aspects of the organization's culture that they would like to modify to allow the company to better adapt to a digital world. They then identify aspects of the desired traits in other organizations and infuse a few teams, a function, or a business unit with the desired characteristics through a series of ongoing minimum viable changes (MVCs). These are changes big enough to help splice in aspects of new digital DNA, but small enough to reduce resistance and rejection. After a series of trial and error experiments, and once these groups begin to exhibit the desired characteristics, then the company begins to propagate the changes demonstrated in different parts of the organization through similar MVC activities.

It may be helpful to return to our discussion of genotype and phenotype from chapter 6. Our argument there is that the core characteristics of good leadership have not changed but simply need to be expressed differently in a new business environment. We make a stronger argument here. Companies need to fundamentally change certain aspects of their organizations to be able to exhibit the characteristics necessary to succeed in the emerging competitive environment. These changes are not simply a case of doing things differently; they are responsive to a need to be different. The only solution is radical intervention to update the essential nature of the organization for a digital world.

The key to accomplishing this goal is to focus first where success is likely to occur, and with one characteristic to change. Our experience and research suggest that this initial focus should be on leaders who are interested and motivated to help the organization become more digital; team members who are capable and interested; and a function, business unit, or team that is already successful in other business endeavors. Inviting these stakeholders will result in more rapid adoption of the change, and like fanned embers of a fire, the change will spread more quickly across the organization. Then, the leaders determine the next organizational characteristic to target for change, and the process is repeated in other similar areas of the company.

Which Digital DNA Traits Are Needed Most by the Organization?

How do an organization's leaders decide which aspects of the organizational DNA to transform using this gene therapy approach? Figure 15.3 shows the twenty-three digital DNA traits, and they are described in greater detail below. As you review the traits, consider the following question: Which of the top three to five digital DNA traits would most move the needle for your organization over the next twelve to eighteen months if those traits were a prominent and mature part of your organization's own DNA?

1. **Continuously innovating:** Because the digital ecosystem is broad, boundaryless, and dynamic, new ideas and different applications of those ideas are constantly needed. Continuously innovating includes developing original and more effective solutions with meaningful impact. This may include products, services, processes, technologies, or business models.

2. **Real time and on demand:** Information, applications, and services are expected by customers, suppliers, partners, and talent to be on demand, up-to-the-moment accurate, accessible around the clock without disruption, and available on multiple platforms and devices.

3. **Shifting decision rights and power:** Decision rights constantly change as valuable information becomes more available to others who previously have not had it, and decision rights change as a result of new processes/work flows. As these decision rights change, the influence of all levels of employees, customers, and other stakeholders are constantly in flux, inside and outside the organization.

4. **Modulating risk and security boundaries:** With digital solutions, there is increased potential for democratization of information, the use of multiple devices, and wider access to information. For these reasons and others, risk and security boundaries are modulated to balance cyber-security requirements with increasing needs for access to information.

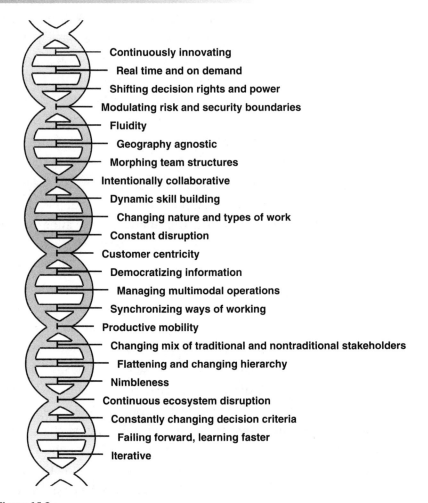

*Evaluate your company based on 23 digital DNA traits
(Pick 3–5 in which to drive change)*

Continuously innovating
Real time and on demand
Shifting decision rights and power
Modulating risk and security boundaries
Fluidity
Geography agnostic
Morphing team structures
Intentionally collaborative
Dynamic skill building
Changing nature and types of work
Constant disruption
Customer centricity
Democratizing information
Managing multimodal operations
Synchronizing ways of working
Productive mobility
Changing mix of traditional and nontraditional stakeholders
Flattening and changing hierarchy
Nimbleness
Continuous ecosystem disruption
Constantly changing decision criteria
Failing forward, learning faster
Iterative

Figure 15.3

5. **Fluidity:** Fluidity is the ability to move with ease from one solution or situation to the next and beyond, while shifts in talent needs, resources, operating models, and communications are occurring. It is planned agility with the capability to execute with a reasonable degree of ease.

6. **Geography agnostic:** Business is becoming less tied to any specific geography. Advances in technology, trends in mobility, and an open talent economy are changing the meaning of "place" or "location."

7. **Morphing team structures:** In digitally maturing environments, teams flexibly form, morph, and disband to meet changing needs. Teams may or may not include members of the organization, customers, partners, vendors, competitors, and others. These teams are designed and sourced with some intentionality and inclusion of diversity of thought and experience.

8. **Intentionally collaborative:** Intentionally collaborating is designing how to work together in sync to carry out common or complementary goals. It is much more than informing or sharing. It is deliberate cooperative design. In digitally maturing organizations, this occurs not only within one's own expected and obvious work groups, but across teams, functions, business units, and, increasingly, beyond the organization.

9. **Dynamic skill building:** Constant innovation, new tasks, and ever-changing ecosystems require flexible skills that can be used to tackle new challenges under new circumstances. It is critical to train talent to develop adaptable skills (e.g., learning to learn, digital literacy, etc.), as well as accommodating flexibility in how and when learning occurs (e.g., "just-enough, just-in-time").

10. **Changing nature and types of work:** Digital disruption and other digital innovations change what work is and how it gets done. Job descriptions, tasks, skills, and requirements are highly fluid in a digital environment. Models of what the work is and how work

gets done (e.g., robots, AI, mixed reality) constantly change in the digital environment.

11. **Constant disruption:** Disruptions in the digital environment are constant and varied. Some are simply noise while others are disintermediating. There are multiple dimensions (e.g., culture, technology, business models, capabilities) that either support or detract from the organization's ability to work in an environment of constant disruption.

12. **Customer centricity:** Customer-centric capabilities put the customer at the center of design thinking and development of products, processes, and decisions. They focus on how to engage, collaborate with, and understand insights from customer interactions, reactions, and aspirations. Customers are often involved in the creation of the goods and services they are using.

13. **Democratizing information:** Customers, the public, suppliers, competitors, employees, contractors, and others increasingly have access to information they may not have had before—and from multiple sources. Digital systems often blur access, availability, security, privacy, and decision rights, from both within the organization and external sources.

14. **Managing multimodal operations:** Multimodal is defined as at least one or more legacy and one or more digital operating models existing within the same organization. This may also include interoperational models with competitors, partners, vendors, and others in the digital ecosystem. Being able to function effectively in different modes of operation simultaneously is a critical capability.

15. **Synchronizing ways of working:** The legacy organization is generally moving at a slower or more inconsistent rate than emerging digital operations. This unevenness can suboptimize digital investments, fail to meet client expectations, or overwhelm the legacy organization. Rewiring processes, expectations, and decision rights at key interaction points between legacy and digital ways of working can increase synchronization and adoption of digital operations.

16. **Productive mobility:** Mobility can make work more convenient, effective, and seamless—irrespective of time, place, device, or medium—and help break down physical and virtual barriers in the workplace. An effective strategy for being productively mobile includes mobile technologies, workstyle, workspaces, and collaboration approaches.

17. **Changing mix of traditional and nontraditional stakeholders:** In a digital, team-based, networked environment, nontraditional stakeholders, who may not formally be a part of an organization's traditional hierarchy, may hold increasingly more power and can affect the success of business outcomes. These traditional and nontraditional people or organizations should be recognized, engaged with, and designed for as stakeholders. Misidentifying, not considering, or ignoring stakeholders could be perilous to an organization's success.

18. **Flattening and changing hierarchy:** Layers of structure, spans of control, and decision rights change rapidly in digital environments. There is generally much less need for layers of structure as work is more technology mediated, teams more networked, and decision flows less hierarchical.

19. **Nimbleness:** The organization is able to adjust to rapid and/or unexpected change. The needed capabilities include speed, skill alignment, flexibility, resourcefulness, and adaptability of systems, processes, people, policies, resources, governance, and so forth.

20. **Continuous ecosystem disruption:** Rapid evolution of how work gets done, where work gets done, and who does the work is disrupting the traditional ecosystem and affecting the network of interactions in which businesses operate. The ability of an organization to operate in this environment of continuous disruption is a key differentiator in this new era of work.

21. **Constantly changing decision criteria:** The speed and complexity of business changes with digital systems. Decision inputs and outputs multiply and shift constantly. What made sense yesterday may not

make sense today or tomorrow because of changes to what customers value, demographic trends, competition, speed of decisions, and interactions with a changing digital ecosystem. There is often unevenness between the digital business/operating models and the legacy organization in how and when decisions are made, and who (people and/or machines) makes them, in this dynamic environment.

22. **Failing forward, learning faster:** Focus is on quickly trying out new or incomplete products, services, or ways of interacting. Teams reflect on what was learned, make rapid adjustments, and try again. This process should emphasize speed, planned agreement of acceptable and unacceptable failure types, and a distinct absence of blame or punishment.

23. **Iterative:** Processes, policies, products, and services are updated and improved based on analytic insights, trial and error, and feedback from development teams, stakeholders, and customers with the intent to get successively closer to a desired result.

Becoming a Digital Organization

When activated, the elements of digital DNA are essential to becoming a digital organization. Creating a digital DNA core for the business and an enabling environment for digital transformation will require commitment and leadership to drive these changes. With digital DNA at the core to help identify areas of focus, organizations will learn, grow, and accelerate their digital ambition to operate successfully in this constantly changing world. Possible areas of focus include:

1. **Work reimagined.** Redesign work activities and processes with the best combination of automation and human skills.

2. **Open talent workforce.** Introduce new talent categories in an open economy that includes salaried, "gig," and ecosystem partners working together seamlessly.

3. **Connected experience.** Deliver an integrated customer and employee experience through physical and virtual workplaces and digital tools that drive productivity and promote personalization.

4. **Digital enablement.** Accelerate business with an insight-driven change analytics and design-thinking approach that puts the customer at the center and embraces iteration.

5. **Network and leadership.** Drive work through an intentionally designed network of teams led by individuals who take appropriate risks and persuade skeptics to deliver the digital ambition.

6. **Digital HR.** Drive the digital evolution for the organization by being digital first and shaping the organization through new cultural behaviors and delivering consumer-grade employee experiences.

In the end, the goal is not to be a digital company for digital's sake but to help create an organization that can operate effectively in the ever-changing future of work.

Enable Change by Leveraging Points of Strength

It would be overwhelming for an organization to try to address twenty-three traits and their corresponding maturity levels simultaneously. Not only is that impractical, it is also not necessary, because which traits and levels of maturity are needed now, as opposed to later, will differ based on an organization's specific needs. Remember that different parts of the organization can also be at different levels of digital maturity. Some more digitally mature functions or business units can help lead the organization toward increased digital maturity through their own success. Eventually, organizations may be able to possess and mature all twenty-three traits, but we have found that in the beginning, it is best to focus on a handful of traits to create transformation capability. In our experience, focusing on three to five digital DNA traits at a time seems to be what most organizations can handle. Ideally, focus on the three to five that will have the most impact in the next twelve to eighteen months. Time bounding leads to more effective prioritization of traits. Other traits will follow because transformation can occur as quickly as the organization is able to absorb changes.

Once leaders have identified the changes from which the organization is most likely to benefit, they often make a critical mistake—designating digital maturity as a "program" rather than an ongoing top-down and bottom-up journey. If this process is treated the way a software implementation would be, it will fail. This isn't about adoption of a new technology; it is about organizing, operating, and behaving effectively in this new world of work.

There are many ways to go about this—some have found that creating a digital function can be a catalyst to becoming a more digitally mature organization; others have been successful with a designated digital leader or leaders; while still others have used a combination of these approaches. Regardless of the approach, getting the embers of a digital fire started and fanning them to spread throughout the organization are critical to adoption. Success begets success. We have learned that digital maturity takes hold best with a top-down *and* a bottom-up approach. Pockets of the organization mature at different rates. Some parts of the organization go after it with gusto, while others cling to traditional ways of operating. We have found three criteria that, when in place, will help spread digital maturity:

1. Identify a digital ambition that is meaningful, rapid, and measurable to achieve.
2. Team up with a sponsoring leader(s) who is already bought in, capable, and motivated to make changes needed.
3. Form teams with members who are capable, open to new approaches, and desirous to participate.

Once the organization has identified the digital traits on which to focus, then it needs to begin assessing how existing aspects of the organization help or hinder infusing the identified traits into the existing organization.

Increase Digital Maturity by Designing and Implementing Small Experimental Interventions

Once you have identified what to change and where to change, then the final step is how to bring that change about. Don't think too long on this point. As we advised in chapter 12, you do not want to adopt large-scale traditional change initiatives, where you try to change the whole organization or even all aspects of the targeted digital DNA trait all at once. Instead, practice what we preach and leverage agile development methods to develop ninety-day (or less) sprints to enact MVCs mentioned earlier in this chapter. These actions are large enough to move your organization toward its vision of digital maturity, but with agile sprint methods to create small enough actions to accelerate the impact. Then, it's critical that you decide how to measure whether the change has been effective, debrief the lessons learned from it, and weave those lessons into ongoing transformation efforts.

Scale and Repeat

The temptation may be to stop there and congratulate yourselves on a successful or attempted intervention sprint (remember, failure *is* an option, as long as you learn something important from it). Understanding the lessons learned from the experiment—whether it is a success or a failure—is important, and those lessons can be used to infuse the learning into the next iteration of experimentation. The key is not to stop with experiments. One minimum viable change does not equal successful digital maturity. Two additional steps are required:

1) **Iterate on the change.** The experiment may have been sufficient at moving your company a little closer to digital maturity. More likely, you'll need to revise the experiment and try it again to improve on the outcomes. Rarely will your experiment get it right the first time, and you need to be continually experimenting and refining that experiment until you have achieved the desired level of change.

2) **Scale the change to other areas of the organization.** If the change has been successful, you can begin driving that change across the organization. We show in chapter 14 that the key differentiator that sets digitally maturing companies apart is not in running these experiments, but in using them to drive change across the organization. Experimentation alone is not enough. Once you have iterated through enough experiments, you need to act on the knowledge you have gained. Indeed, here we return to one of the opening sections

of the book: "Thinking is for doing." Use the experiments to narrow the knowing-doing gap.

We hope this process will help your organization quickly realize that change is possible, even if only in small pockets at first. The key is to keep focusing on helping those small pockets mature and supporting them in the process. As you begin to rack up small wins, other parts of the organization will begin to buy in, and the pace of transformation can accelerate rapidly. This practical process results in your organization's DNA being infused with new (recombinant) DNA, with greater levels of the digital maturity needed to thrive in this constantly changing future of work.

Conclusion: There's No Going Back to Kansas

According to the Library of Congress, *The Wizard of Oz* is the most watched film ever in the United States.[1] If you are one of the millions of people who have watched the movie since its 1939 cinematic debut, then you know that the film concludes with Dorothy's awakening in her home in Kansas and uttering the memorable words, "There's no place like home."[2] In the movie, Dorothy's entire transformational journey through Oz turns out to be nothing more than a dream. Many people may similarly hope that the need for the digital transformation journey is a dream. Like Judy Garland's Dorothy Gale, they hope that they can return home to the world as it was before the cyclone of digital disruption.

There are many differences between the book and the film—in the book, for example, Dorothy has silver shoes, not ruby slippers; the color of the shoes was changed in the movie to take advantage of the new Technicolor film process. But the most crucial difference is that in Baum's books, Dorothy's visit to Oz is not a dream—she actually goes there. As we have tried to make clear throughout this book, we believe that digital disruption is here to stay. What we see today is only part of a series of long-term disruptions that we expect to intensify in the coming decades. Remember that in the book, when Dorothy returns to Kansas, she finds it hopelessly dull and boring after her exciting journey through Oz. She doesn't *want* to go back to her old life. She likes the new adventure better.

We've seen a similar theme play out repeatedly when we present our research on digital maturity. People who have been employed by organizations that have gone through the process of working toward digital maturity invariably describe it as painful. The change, the disruption, and having to learn new ways of working all require considerable effort. Complaints and failures inevitably occur along the way. The transformation effort can be just as unsettling as finding oneself whisked away to Oz by a cyclone. Nevertheless, most of these individuals, once they were well along on the road to digital transformation, also reported that they would never want to go back to the "old way." Whether they are drawn in by the medical doctor using an innovative digital diagnostic tool or by a professional using advanced digital collaboration tools, once they make the transition to more digitally mature ways of working, they recognize how much better and more productive work in this new world can be. Just as most of us would never voluntarily go back to the days before smartphones and personal computers, these people don't want to go back to digitally immature organizations.

Employees who have chosen to go back to "the old way," often by joining less mature companies to secure better positions, frequently comment on how frustrating the experience is. They know that there are better and more efficient ways to do things, but their new organization resists the types of changes and ways of working that they know are possible. They feel like they are taking big steps backward in productivity and possibility.

There's No Going Back. . . .

This realization leads to a final implication from our extended metaphor. The story of digital disruption that we cover in this book may only be one chapter among many more to come. To be fair, at this point, we may already be well into the third or fourth chapter of digital disruption—from mainframes in the 1960s and 1970s, to personal computers in the 1980s through the mid-1990s, to the internet in the late 1990s into the 2000s, to the mobile and analytics revolutions disrupting businesses today. Nevertheless, wherever we may find ourselves in the digital disruption continuum, the point remains the same—the story of digital disruption is far from over, and there are many more chapters to come.

Digital disruption won't end anytime soon. As new technologies make their way into the mainstream—blockchain, AI, autonomous vehicles, additive manufacturing, virtual and augmented reality—businesses will face new questions about how to use technologies to run their businesses differently. Inevitably, another set of technological advances that we cannot even envision today will follow. Considering that the pace of technological advancement continues growing exponentially, it is highly likely that the average worker will experience multiple waves of digital disruption before his or her career ends.

While we hope that the key lessons of this book help executives deal with digital disruption today, we also believe that many of the insights will be valuable through multiple future waves of disruption. For example:

- In part I, we argue that digital disruption is fundamentally about people, which we expect to be true in the future disruptions too. Individuals, organizations, and societies will continue to adapt to technological change at different rates, re-creating similar problems in the future as organizations experience today. Business leaders will continue to need to strategize amid an uncertain and rapidly changing future. Companies will still need to continually mature—adapting themselves to act in ways that are appropriate for the new reality brought by digital disruption.

- In part II, we discuss the nature of talent and leadership in a digital business environment. Many of these lessons will remain true as well. A growth mindset will likely be important in future waves of digital disruption. People will continue to want to work for digitally maturing organizations—whatever those happen to look like in the future—and organizations will want to attract and retain the most talented workers. We expect that certain essential leadership traits will continue to persist in the future, yet how leaders express those traits will continue to evolve as a result of the environment in which they are expressed. Work will likely look very different in the future, but we expect that work will continue to exist.

- In part III, we discuss the organizational characteristics essential for digital maturity. This section may be the most tailored to the current digital environment, and therefore it has the most subject to change in future waves of disruption. We certainly expect that the need to experiment, take an appropriate amount of risk, and collaborate will remain important in future waves of digital disruption. How those experiments will be conducted, what will be considered an appropriate amount of risk, and how and with whom/with what will we collaborate are questions whose answers may look considerably different from how digitally maturing companies look today.

Continual Learning as the Best Response to Digital Disruption

The danger of writing a book about digital transformation is the potential for (rapid) obsolescence. What is new and novel today can easily become outdated and passé tomorrow. What may have once seemed impossible can, in retrospect, seem inevitable.[3] Consequently, as we were casting about for the focus of this book, we concluded that we would never outrun the threat of obsolescence. So, we opted to concentrate this book on the underlying organizational challenges that will likely remain relevant in future waves of disruption. Nevertheless, even with this focus, parts of this book will—by definition—become obsolete over time. When we teach about digital disruption in the business school, we make it clear to students that any of the practical guidance we provide in the class is almost surely to be outdated in five years.

One does not need to be enrolled in a university to continue individual digital education. The abundance of online digital communities geared toward developing digital skills also makes this the golden age of continual learning. Sites like Code Academy can help you learn programming languages. Numerous massive online open courses (MOOCs) provide deeper dives into topics like machine learning, R programming, and data science (interestingly, one of the most popular MOOCs is about learning how to learn).[4] TED talks deal with cutting-edge topics and are often deeper and more insightful than much of the short-form content being produced by many traditional business publications. These talks are publicly available to all. You can follow thought leaders directly on Twitter, which often provides important insight and connection to genuine experts. These insights can go beyond those associated with digital disruption. For example, we once used Twitter to get clarification about a turkey recipe from *Top Chef* star Tom Colicchio on Thanksgiving Day.

While the need to engage in continual learning is stronger than ever before, the opportunities for doing so are also much greater than ever before. The challenge, of course, is first developing sufficient digital literacy to be able to distinguish the genuine experts from the self-promoting snake oil salesmen, who are equally abundant online.

Maintaining an Organizational Growth Mindset

Continual learning is also needed at the organizational level. We introduce the concept of an individual-level growth mindset in chapter 8, but some scholars have posited that we need an organizational-level growth mindset as well. Just as individuals need to recognize that certain traits are obtained or strengthened as a result of hard work, organizations also need to recognize that they can obtain the traits of a digitally maturing organization through similar hard work.

Organizations with a growth mindset emphasize learning *as individuals* and *as organizations*, and studies have linked these organizations with increased innovation, collaboration, and risk taking.[5] Carol Dweck points out three common *misconceptions* about a growth mindset that also describe traps that organizations fall into when trying to mature digitally.[6]

1. **I already have it, and I always have.** Dweck notes that open mindedness is not equivalent to a growth mindset. "Individuals who believe their talents can be developed (through hard work, good strategies, and input from others) have a growth mindset. . . . People often confuse a growth mindset with being flexible or open-minded or with having a positive outlook." Being open minded or having a

positive outlook is helpful for learning, as it makes the person receptive to new information and perspectives. But it's not sufficient. A growth mindset must include embracing challenges, perseverance, and the kind of hard work and effort that results in learning and mastery. Cultivating a growth mindset needs to be a constant effort and is particularly important in a world of digital disruption.

2. **A growth mindset is just about praising and rewarding effort.** Dweck notes that experimentation just for the sake of experimentation is not particularly helpful. Experimentation, feedback, and iteration must aim at moving toward tangible goals. That's why we are clear to add the last part to "test fast, learn fast, scale fast." Trying is easy. Failure is easy. Productive learning that moves your organization toward a desired goal is not.

3. **Just espouse a growth mindset, and good things will happen.** Individuals cannot obtain a growth mindset just by claiming they have it; they must act accordingly. Likewise, organizations cannot just claim to be digital, agile, or risk tolerant and expect to be so; they need to act that way as well. Yet, our data suggest that this gap between what organizations say and how they act is common when it comes to digital transformation. We see a big split in our data between the rosier perspective espoused by leaders and the more pessimistic view of employees. Growth mindset is about acting differently as an organization.

Organizations can and should work at developing a growth mindset, as it is a key facet of digital maturity.

Creating Organizational Digital Literacy

Digital literacy is the second key concept that should be extended to the organizational level to help engender digital maturity. While individual digital maturity remains important, organizations need their employees to have some level of base common knowledge to enable productive communication and collaboration. Organizational digital literacy is likely better defined by the lowest level of digital literacy among employees, rather than its average. Like our congruence concept in chapter 3, a small number of employees who are unable or unwilling to work in digitally maturing ways can slow down the entire organization. Several of our respondents noted the necessity of parting ways with employees who were unwilling to continue developing their digital skills to meet business needs.

While you certainly don't want your employees to all have the same knowledge, lest there is no reason for them to work together, neither do you want them to all have completely different knowledge, lest they will be unable to communicate with one another. To be clear, digital literacy is *not* the occasional IT training course, which often has little real effect on employee performance. Instead it's about helping employees begin to think about their work differently in a digital world. How can employees be expected to experiment and iterate if they don't have a basic working knowledge of the underlying phenomena to which they are trying to adapt? Individuals may indeed adapt to technology faster than organizations do, but they do so in ways that benefit them as individuals. Adapting in a way that benefits their organization is more challenging.

Many organizations do little to encourage their employees to obtain this basic digital literacy, which may explain some of the frustration employees feel about their organization's support for developing digital skills. Furthermore, organizations need to give employees time to become more digitally literate. One employee notes, "Our CEO is on Chatter a lot, but I'm working 10–12 hours per day just to get my job done[;] I don't have time to check that, too." Freeing up time to enable employees to expand their digital skills may make the other hours that they spend working far more productive.

The bright side is that organizational digital literacy is readily attainable with a little bit of effort. Academic research suggests that a little bit of new knowledge can often have a substantial effect on learning outcomes.[7] A little bit can go a long way, but many organizations provide little opportunity for their employees to obtain even minimally sufficient amounts. If your organization does seek to implement formal initiatives to expand your employee's digital literature, a combination of online and offline learning approaches often yields the best results. Colocated learning often builds relationships and allows more productive dialogue than online settings alone. Yet, limiting learning to in-person settings alone is far too limited to be productive. People coming together for a few hours every month or every quarter is unlikely to have a meaningful influence on your organization's digital literacy. Instead, the best environments combine online and offline engagement in ways that mutually reinforce each other. Offline sessions typically allow deeper engagement and reflection for the content generated online, and online engagement typically continues the offline discussions. If you can get one of these online communities going, it can have significant implications for your organization's digital literacy and, consequently, its digital maturity.

This Is the End. . . .

Thanks for joining us on this journey to explore how organizations respond to digital disruption. This book is not meant as a recipe, and many of the digitally maturing organizations we profiled in our annual surveys would have a hard time describing exactly how they got to where they find themselves today. We have tried to identify the organizational characteristics that are common to digitally maturing organizations and suggest what organizations might do to cultivate, nurture, and enhance these characteristics. Our goal is to demystify what it means to mature digitally and to help organizations emphasize the characteristics that align with this state.

The key to getting started is not necessarily to push the most mature parts of your organization further, but instead to begin knocking down the barriers to digital maturity. You may find that in doing so, you empower other areas of your organization that were just waiting to move. Once you begin to make progress, your company is likely to gain momentum, making subsequent efforts easier. Starting is often the hardest part. Good luck and send us stories of your successes, large and small, as well as the lessons you learn along the way!

Notes

Introduction

1. The reports this book is based on are by G. C. Kane, D. Palmer, A. N. Phillips, D. Kiron, and N. Buckley: "Strategy, Not Technology, Drives Digital Transformation," *MIT SMR* Report on Digital Business, July 14, 2015, https://sloanreview.mit.edu/projects/strategy-drives-digital-transformation/; "Aligning the Organization for its Digital Future," *MIT SMR* Report on Digital Business, July 26, 2016, https://sloanreview.mit.edu/projects/aligning-for-digital-future/; "Achieving Digital Maturity," *MIT SMR* Report on Digital Business, July 13, 2017, https://sloanreview.mit.edu/projects/achieving-digital-maturity/; and "Coming of Age Digitally: Learning, Leadership, and Legacy," *MIT SMR*/Deloitte Report on Digital Business, June 5, 2018, https://sloanreview.mit.edu/projects/coming-of-age-digitally/.

Chapter 1

1. Jeffrey Pfeffer and Robert I. Sutton, *The Knowing Doing Gap* (Boston: Harvard Business School Press, 1999), 1, 246.

2. Kane et al., "Coming of Age Digitally."

3. Walt Kelly Earth Day poster, April 22, 1970, cited in This Day in Quotes, April 22, 2015, http://www.thisdayinquotes.com/2011/04/we-have-met-enemy-and-he-is-us.html.

4. A. Abbatiello, D. Agarwal, J. Bersin, G. Lahiri, J. Schwartz, and E. Volini, "The Workforce Ecosystem: Managing beyond the Enterprise," 2018 Global Human Capital Trends report, Deloitte Insights, March 28, 2018, https://hctrendsapp.deloitte.com/reports/2018/the-workforce-ecosystem.html.

Chapter 2

1. Gerald C. Kane, "Digital Disruption Is a People Problem," *MIT SMR*, September 18, 2017, https://sloanreview.mit.edu/article/digital-disruption-is-a-people-problem/.

2. Everett M. Rogers, *Diffusion of Innovations* (New York: Free Press, 1962).

3. The technology adoption curve is well known and has been the subject of numerous derivative works. One of the better-known such works is *Crossing the Chasm*, by Geoffrey A. Moore (1991; rev. ed., New York: Harper Business Essentials, 2014). Moore argues that the adoption process is discontinuous, with a significant gap between early adopters and the mass market.

4. Kenneth Kiesnoski, "The Top 10 US Companies by Market Capitalization," CNBC, last updated October 24, 2017, https://www.cnbc.com/2017/03/08/the-top-10-us-companies-by-market-capitalization.html.

5. Ron Kohavi and Stefan Thomke, "The Surprising Power of Online Experiments," *Harvard Business Review* 95, no. 5 (September-October 2017): 74.

6. Andrea Huspeni, "Why Mark Zuckerberg Runs 10,000 Facebook Versions a Day," *Entrepreneur*, accessed July 25, 2017, https://www.entrepreneur.com/article/294242.

7. Oliver E. Williamson, *Markets and Hierarchies: Analysis and Antitrust Implications, a Study of the Economics of International Organization* (New York: Free Press, 1975).

8. Robert M. Grant, "Toward a Knowledge-Based Theory of the Firm," *Strategic Management Journal* 17 (1996): 109–122.

9. Wesley M. Cohen and Daniel A. Levinthal, "Absorptive Capacity: A New Perspective on Learning and Innovation," *Administrative Science Quarterly* 35, no. 1 (1990): 126, 128.

10. Ibid., 131.

11. Shaker Zahra and Gerard George, "Absorptive Capacity: A Review, Reconceptualization, and Extension," *Academy of Management Review* 27, no. 2 (2002): 182–203.

Chapter 3

1. David A. Nadler and Michael L. Tushman, "A Model for Diagnosing Organizational Behavior," *Organizational Dynamics* 9, no. 2 (1980): 35–51.

2. Rita Rani Talukdar and Joysree Das, "A Study on Emotional Maturity Among Arranged Marriage Couples," *International Journal of Humanities and Social Science Invention* 2, no. 8 (August 2013): 16–18.

3. In the Supreme Court case *Jacobellis v. Ohio*, Stewart wrote, "I shall not today attempt further to define the kinds of material I understand to be embraced within that shorthand description [hardcore pornography], and perhaps I could never succeed in intelligibly doing so. But *I know it when I see it*, and the motion picture involved in this case is not that." See more at "Jacobellis v. Ohio," Legal Information Institute, Cornell Law School, accessed October 15, 2017, https://www.law.cornell.edu/supremecourt/text/378/184.

4. *The Start-Up of You* website, accessed October 15, 2017, http://www.thestartupofyou.com/.

Chapter 4

1. *Leading Digital* website, accessed October 25, 2017, http://www.leadingdigitalbook.com/.

2. *Machine, Platform, Crowd* website, accessed October 25, 2017, http://books.wwnorton.com/books/Machine-Platform-Crowd/.

3. "Books and Research," DavidRogers.biz, accessed October 25, 2017, http://www.davidrogers.biz/books-research/.

4. *Platform Revolution* website, accessed October 25, 2017, http://books.wwnorton.com/books/detail.aspx?ID=4294993559.

5. John Gallaugher's website, accessed October 25, 2017, http://gallaugher.com/book/.

6. For a cynical look at this phenomenon, read "Good to Great to Gone," *Economist*, July 7, 2009, http://www.economist.com/node/13980976.

7. James G. March, "Exploration and Exploitation in Organizational Learning," *Organization Science* 2, no. 1 (1991): 71–87.

8. Charles A. O'Reilly and Michael L. Tushman, "The Ambidextrous Organization," *Harvard Business Review* 82, no. 4 (2004): 74–81.

9. John Hagel and John Seely Brown, "Zoom Out/Zoom In: An Alternative Approach to Strategy in a World That Defies Prediction," Deloitte Insights, May 16, 2018, https://www2.deloitte.com/insights/us/en/topics/strategy/alternative -approach-to-building-a-strategic-plan-businesses.html.

10. Gerald C. Kane, "Predicting the Future: How to Engage in Really Long-Term Strategic Digital Planning," *Big Idea: Digital Leadership* (blog), *MIT SMR*, May 3, 2016, https://sloanreview.mit.edu/article/predicting-the-future-how-to-engage -in-really-long-term-strategic-digital-planning/.

11. "Olo's Dispatch Brings On-Demand Delivery Service to Restaurants and Consumers Nationwide: Integrated Handoff to Delivery Partners Yields Transit Times under 700 Seconds on Average," Business Wire, September 14, 2016, https:// www.businesswire.com/news/home/20160914005244/en/Olo%E2%80%99s -Dispatch-Brings-On-Demand-Delivery-Service-Restaurants.

12. Supantha Mukherjee, "Amazon Is Gearing up to Deliver Food from Restaurants like Chipotle, Applebee's, and Shake Shack," *Business Insider*, September 22, 2017, http://www.businessinsider.com/r-food-ordering-company-olo -ties-up-with-amazon-restaurants-2017-9.

Chapter 5

1. James J. Gibson, *The Ecological Approach to Visual Perception* (Boston: Houghton Mifflin Harcourt, 1979), 127.

2. Carliss Baldwin and Kim Clark, *Design Rules*, vol. 1, *The Power of Modularity* (Cambridge, MA: MIT Press, 2000).

3. David M. Ewalt, "The Other Greatest Tool Ever," *Forbes*, March 15, 2006, https://www.forbes.com/2006/03/14/tools-duct-tape_cx_de_0315ducttape .html.

4. Amazon search results for "duct tape" in the Books section, accessed October 25, 2017, https://www.amazon.com/s/ref=nb_sb_noss_2?url=search -alias%3Dstripbooks&field-keywords=duct+tape&rh=n%3A283155%2Ck%3A duct+tape.

5. @RKRosengard, "Tips for Engaging Live: How Automakers Used Periscope at #NYIAS," Twitter Marketing (blog), March 31, 2016, https://blog.twitter.com /marketing/en_us/a/2016/tips-for-engaging-live-how-automakers-used-peri scope-at-nyias.html.

6. Jamie Moore, "What It Means to Be a Black Belt," International Taekwon-Do Federation, April 12, 2006, http://www.itf-administration.com/articles.asp?arturn=668.

7. Matt Rocheleau, "The Pedestrian Buttons at Crosswalks? They Don't Actually Do Anything," *Boston Globe*, July 24, 2017.

8. David McRaney, "Placebo Buttons," *You Are Not So Smart*, February 10, 2010, https://youarenotsosmart.com/2010/02/10/placebo-buttons/.

9. John Gallaugher, *Information Systems: A Manager's Guide to Harnessing Technology, v. 6.0* (Boston: FlatWorld, 2017).

10. Ibid.

11. Paul M. Leonardi, "When Does Technology Use Enable Network Change in Organizations? A Comparative Study of Feature Use and Shared Affordances," *MIS Quarterly* 37, no. 3 (2013): 749–775.

12. For additional information on Leonardi and his work, see his faculty bio on the Technology Management Program website, UC Santa Barbara, accessed October 25, 2017, https://tmp.ucsb.edu/about/people/paul-leonardi/home.

Chapter 6

1. Guatam Mukunda, "Jefferson and Lincoln: Different Leaders for Different Times," *Fortune*, February 18, 2013, http://fortune.com/2013/02/18/jefferson-and-lincoln-different-leaders-for-different-times/.

2. John Hagel, John Seely Brown, Andrew de Maar, and Maggie Wooll, "Beyond Process: How to Get Better, Faster as 'Exceptions' Become the Rule," Deloitte Insights, November 13, 2017, https://www2.deloitte.com/insights/us/en/topics/talent/business-process-redesign-performance-improvement.html.

3. A good explanation of Johannsen's work is B. R. Erick Peirson, "Wilhelm Johannsen's Genotype-Phenotype Distinction," *The Embryo Project Encyclopedia*, December 7, 2012, http://embryo.asu.edu/pages/wilhelm-johannsens-genotype-phenotype-distinction.

4. S. R. Barley, "The Alignment of Technology and Structure through Roles and Networks," *Administrative Science Quarterly* 35, no. 1 (1990): 61–103.

5. Arthur C. Clarke, *Profiles of the Future: An Inquiry into the Limits of the Possible* (New York: Popular Library, 1973).

Chapter 7

1. Shamel Addas, Alain Pinsonneault, and Gerald C. Kane, "Converting Email from a Drain into a Gain," *MIT SMR* 59, no. 4 (Summer 2018), https://sloanre view.mit.edu/article/converting-email-from-drain-to-gain/.

2. Robert D. Austin and David M. Upton, "Leading in the Age of Super-Transparency," *MIT SMR* 57, no. 2 (Winter 2016), https://sloanreview.mit.edu /article/leading-in-the-age-of-super-transparency/.

3. Kane et al., "Coming of Age Digitally."

Chapter 8

1. For a good discussion of this issue, see Anthony J. Bradley and Mark P. McDonald, "People Are Not Your Greatest Asset," *Harvard Business Review*, December 6, 2011, https://hbr.org/2011/12/people-are-not-your-greatest-a.

2. Larry Abramson, "Sputnik Left Legacy for U.S. Science Education," NPR, September 30, 2007, http://www.npr.org/templates/story/story.php?storyId=14829195.

3. Dennis Vilorio, "STEM 101: Intro to Tomorrow's Jobs," *Occupational Outlook Quarterly*, Spring 2014, 3–12.

4. Carol S. Dweck, *Mindset: The New Psychology of Success* (New York: Ballantine Books, 2006).

5. G. C. Kane, D. Palmer, A. Phillips, and D. Kiron, "Is Your Business Ready for a Digital Future?" *MIT SMR* 56, no. 4 (Summer 2015). http://staging.mitsmr.io /article/is-your-business-ready-for-a-digital-future/.

6. "How Companies Can Profit from a 'Growth Mindset,'" *Harvard Business Review*, November 2014, https://hbr.org/2014/11/how-companies-can-profit-from -a-growth-mindset.

7. Ibid.

8. Ibid.

9. K. Monahan, T. Murphy, and M. Johnson, "Humanizing Change: Developing More Effective Change Management Strategies," *Deloitte Review,* no. 19 (July 14, 2016), https://www2.deloitte.com/insights/us/en/deloitte-review/issue-19/devel oping-more-effective-change-management-strategies.html.

Chapter 9

Epigraph: "About Chez Panisse," Chez Panisse Restaurant and Café website, accessed January 19, 2018, http://www.chezpanisse.com/about/chez-panisse/.

1. "Chez Panisse," *Wikipedia*, last edited August 5, 2018, https://en.wikipedia.org/wiki/Chez_Panisse.

2. *Superbosses: How Exceptional Leaders Master the Flow of Talent*, Tuck School of Business at Dartmouth, accessed January 19, 2018, http://www.superbosses.com/.

3. "Olo's Dispatch Brings On-Demand Delivery Service to Restaurants and Consumers Nationwide: Integrated Handoff to Delivery Partners Yields Transit Times under 700 Seconds on Average," Business Wire, September 14, 2016, https://www.businesswire.com/news/home/20160914005244/en/Olo%E2%80%99s-Dispatch-Brings-On-Demand-Delivery-Service-Restaurants.

4. Thomas Peters and Robert H. Waterman Jr., *In Search of Excellence* (New York: HarperCollins, 2004), 289.

5. Anne Fisher, "Management by Walking Around: 6 Tips to Make It Work," *Fortune*, August 23, 2012, http://fortune.com/2012/08/23/management-by-walking-around-6-tips-to-make-it-work/.

Chapter 10

1. "Luddite," *Urban Dictionary*, April 24, 2004, https://www.urbandictionary.com/define.php?term=luddite.

2. Richard Conniff, "What the Luddites Really Fought Against," *Smithsonian Magazine*, March 2011, https://www.smithsonianmag.com/history/what-the-luddites-really-fought-against-264412/.

3. Ibid.

4. Arwa Mahdawi, "What Jobs Will Still Be Around in 20 Years?" *Guardian*, June 26, 2017, https://www.theguardian.com/us-news/2017/jun/26/jobs-future-automation-robots-skills-creative-health.

5. David Autor, "Will Automation Take Away All Our Jobs?" TEDxCambridge, September 2016, video, 18:38, https://www.ted.com/talks/david_autor_why_are_there_still_so_many_jobs.

6. Ian D. Wyatt and Daniel E. Hecker, "Occupational Changes during the 20th Century," *Monthly Labor Review*, March 2006, https://www.bls.gov/mlr/2006/03/art3full.pdf.

7. Reid Wilson, "Census: More Americans Have College Degrees than Ever Before," Hill, April 3, 2017, http://thehill.com/homenews/state-watch/326995-census-more-americans-have-college-degrees-than-ever-before.

8. Derek Thompson, "A World Without Work," *Atlantic*, July–August 2015, 50–61.

9. Cathy Engelbert and John Hagel, "Radically Open: Tom Friedman on Jobs, Learning, and the Future of Work," *Deloitte Review*, no. 21 (July 2017): 105.

10. Anthony Goldbloom, "The Jobs We'll Lose to Machines—and the Ones We Won't," TED2016, February 2016, video, 4:37, https://www.ted.com/talks/anthony_goldbloom_the_jobs_we_ll_lose_to_machines_and_the_ones_we_won_t.

11. "The Computer Will See You Now," *Economist*, August 20, 2014, https://www.economist.com/news/science-and-technology/21612114-virtual-shrink-may-sometimes-be-better-real-thing-computer-will-see.

12. Elizabeth Gibney, "Self-Taught AI Is Best Yet at Strategy Game Go," *Nature News*, October 18, 2017, https://www.nature.com/news/self-taught-ai-is-best-yet-at-strategy-game-go-1.22858.

13. Marco Iansiti and Karim R. Lakhani, "The Truth about Blockchain," *Harvard Business Review*, January–February 2017, https://hbr.org/2017/01/the-truth-about-blockchain.

14. Josh Bersin, "Catch the Wave: The 21st-Century Career," *Deloitte Review*, no. 21 (July 2017), https://www2.deloitte.com/insights/us/en/deloitte-review/issue-21/changing-nature-of-careers-in-21st-century.html.

15. Thomas H. Davenport and Julia Kirby, "Beyond Automation," *Harvard Business Review*, June 2015, https://hbr.org/2015/06/beyond-automation.

16. Conniff, "What the Luddites Really Fought Against."

Chapter 11

1. There is some debate whether Drucker ever actually said this or its equivalent. As Andrew Cave reports in a column for *Forbes*, "His alleged phrase, which doesn't seem to appear in any of his thirty-nine books, has passed into corporate-speak and is oft-referred to by management consultants." In fact, "the phrase was

attributed to him in 2006 by Mark Fields, who later became chief executive of motor giant Ford, and eleven years on it is gaining new currency as part of the movement for purpose-led business." For additional commentary, see Andrew Cave, "Culture Eats Strategy for Breakfast. So What's for Lunch?" *Forbes*, November 9, 2017, https://www.forbes.com/sites/andrewcave/2017/11/09/cul ture-eats-strategy-for-breakfast-so-whats-for-lunch/#5e12d3dd7e0f.

2. Schein's biography and research can be found at "Faculty and Research," MIT Sloan School of Management, accessed December 13, 2017, http://mitsloan .mit.edu/faculty-and-research/faculty-directory/detail/?id=41040.

3. This description of Schein's work, from "Coming to a New Awareness of Organizational Culture," in *Behavior in Organizations*, ed. J. B. Lau and A. B. Shani (Homewood, IL: Irwin, 1988), 375–390, appears in slides by Mike Hoseus, Lean Culture Enterprises, accessed December 13, 2017, http://www.ccmm.ca/docu ments/formationContinue/presentations/2007_2008/08_02_13_ToyotaWay -presentation.pdf.

4. For additional information, see *The Best Place to Work* website, accessed December 13, 2017, http://thebestplacetoworkbook.com/.

5. See Peirson, "Wilhelm Johannsen's Genotype-Phenotype Distinction."

6. L. Lapointe and S. Rivard, "A Multilevel Model of Resistance to Information Technology Implementation," *MIS Quarterly* 29, no. 3 (2005): 461–491.

7. Attributed to Truman by Richard E. Neustadt, *Presidential Power: The Politics of Leadership* (New York: John Wiley, 1960), 9, cited by Bartleby.com, accessed December 13, 2017, http://www.bartleby.com/73/1514.html.

8. "*Field of Dreams* (1989)," Internet Movie Database, accessed December 13, 2017, http://www.imdb.com/title/tt0097351/.

9. Lynn Wu and Gerald Kane, "Network-Biased Technical Change: How Social Media Tools Disproportionately Affect Employee Performance," *SSRN*, March 7, 2016, https://papers.ssrn.com/sol3/papers.cfm?abstract_id=2433113.

10. Dion Hinchcliffe, "Enterprise 2.0 success: BASF," *ZDNet*, February 15, 2012, http://www.zdnet.com/article/enterprise-2-0-success-basf/.

11. Yochai Benkler, "The Unselfish Gene," *Harvard Business Review*, July–August 2011, https://hbr.org/2011/07/the-unselfish-gene.

12. See "Cluster Analysis," *Wikipedia*, last updated August 7, 2018, https:// en.wikipedia.org/wiki/Cluster_analysis, for a further description of cluster analysis as a statistical technique.

13. The persistence of our three-cluster result called forth memories from *Monty Python and the Holy Grail*: "Then, shalt thou count to three. No more. No less. Three shalt be the number thou shalt count, and the number of the counting shall be three. Four shalt thou not count, nor either count thou two, excepting that thou then proceed to three. Five is right out." Cited by Board of Wisdom, accessed December 13, 2017, https://boardofwisdom.com/togo /Quotes/ShowQuote?msgid=6866#.WjHm4UxFzD4.

14. For a discussion of this phenomenon, see Alex Kaufman, "Prepare to Be Shocked! What Happens When You Actually Click on One of Those 'One Weird Trick' Ads?" *Slate*, July 30, 2013, http://www.slate.com/articles/business/money box/2013/07/how_one_weird_trick_conquered_the_internet_what_happens _when_you_click_on.html.

Chapter 12

1. You can read more at "Manifesto for Agile Software Development," Agile Alliance, February 2001, https://www.agilealliance.org/agile101/the-agile -manifesto/.

2. Julian Birkinshaw, "What to Expect from Agile," *MIT SMR* 59, no. 2 (Winter 2018), https://sloanreview.mit.edu/article/what-to-expect-from-agile/.

3. Steve Denning, "What Is Agile?" *Forbes*, August 13, 2016, https://www.forbes .com/sites/stevedenning/2016/08/13/what-is-agile/#4d1e081326e3.

4. Steve Banker, "3D Printing Revolutionizes the Hearing Aid Business," *Forbes*, October 15, 2013, https://www.forbes.com/sites/stevebanker/2013/10/15 /3d-printing-revolutionizes-the-hearing-aid-business.

5. Ibid.

6. See, for example, Rachel Gillett, "Productivity Hack of the Week: The Two Pizza Approach to Productive Teamwork," *Fast Company*, October 24, 2014, https://www.fastcompany.com/3037542/productivity-hack-of-the-week-the -two-pizza-approach-to-productive-teamwork.

7. For a complete discussion of this quotation and its context, see "No Plan Survives Contact with the Enemy," Boot Camp and Military Fitness Institute, February 28, 2016, https://bootcampmilitaryfitnessinstitute.com/military-and-outdoor-fitness -articles/no-plan-survives-contact-with-the-enemy/.

8. Birkinshaw, "What to Expect from Agile."

9. Carliss Y. Baldwin and Kim B. Clark, *Design Rules, Volume 1: The Power of Modularity* (Cambridge, MA: MIT Press, 2000).

10. Karl E. Weick, "Educational Organizations as Loosely Coupled Systems," *Administrative Science Quarterly* 21 (1976): 1–19.

11. G. C. Kane, D. Palmer, A. N. Phillips, and D. Kiron, "Winning the Digital War for Talent." *MIT SMR* 58, no. 2 (Winter 2017), https://sloanreview.mit.edu/article/winning-the-digital-war-for-talent/.

12. Birkinshaw, "What to Expect from Agile."

Chapter 13

1. Katie Stagg, "Catalonia's Human Towers: The Art of Castells," *Culture Trip*, updated April 25, 2018, https://theculturetrip.com/europe/spain/articles/catalonia-s-human-towers-the-art-of-castells/.

2. "Castell (Human Tower) Performance in Barcelona, Spain," *HitchHikers-Handbook.com*, May 18, 2013, http://hitchhikershandbook.com/2013/05/18/castell-human-tower-performance-in-barcelona-spain.

3. "Human Towers," Intangible Cultural Heritage, UNESCO, accessed January 4, 2018, https://ich.unesco.org/en/RL/human-towers-00364.

4. Ibid.

5. For example, "Barcelona Human Tower (Castell)," YouTube, video, 4:20, uploaded July 30, 2013, by Leópold Kristjánsson, https://www.youtube.com/watch?v=UnCi_a7P-WM.

6. "Castell (Human Tower) Performance in Barcelona, Spain." Microsoft Translator translates *seny* as either "sanity" or "wisdom."

7. Wu and Kane, "Network-Biased Technical Change."

8. "What Is Sociometry," Toronto Centre for Psychodrama and Sociometry, accessed January 4, 2018, http://www.tcps.on.ca/learn/sociometry.

9. Gerald C. Kane, "Digital Transparency and Permanence," *Big Idea: Social Business* (blog), October 6, 2015, https://sloanreview.mit.edu/article/digital-transparency-and-permanence/.

10. Burcu Bulgurcu, Wietske Van Osch, and Gerald C. Kane, "The Rise of the Promoters: User Classes and Contribution Patterns in Enterprise Social Media," *Journal of Management Information Systems* 35, no. 2 (2018): 610–646.

11. Yuqing Ren and Linda Argote, "Transactive Memory Systems 1985–2010: An Integrative Framework of Key Dimensions, Antecedents, and Consequences," *Academy of Management Annals* 5, no. 1 (June 2011): 189–229.

12. Irving L. Janis, *Victims of Groupthink* (Boston: Houghton Mifflin, 1972).

13. James Surowecki, *The Wisdom of Crowds* (New York: Doubleday, 2004).

14. The MIT Center for Collective Intelligence, accessed January 4, 2018, http:// cci.mit.edu/index.html.

15. Thomas W. Malone and Michael S. Bernstein, eds., *The Handbook of Collective Intelligence* (Cambridge, MA: MIT Press, 2015).

16. Quy Huy and Andrew Shipilov, "The Key to Social Media Success in Organizations," *MIT SMR* 54, no. 1 (Fall 2012), https://sloanreview.mit.edu/article /the-key-to-social-media-success-within-organizations/.

Chapter 14

1. See, for example, Dominic Basulto, "The New #Fail: Fail Fast, Fail Early and Fail Often," *Washington Post*, May 30, 2012, https://www.washingtonpost.com /blogs/innovations/post/the-new-fail-fail-fast-fail-early-and-fail-often/2012 /05/30/gJQAKA891U_blog.html.

2. Ibid.

3. For an interesting perspective on whether "fail fast" is an accurate way to describe innovation, see Rob Asghar, "Why Silicon Valley's 'Fail Fast' Mantra Is Just Hype," *Forbes*, July 14, 2014, https://www.forbes.com/sites/robasghar/2014 /07/14/why-silicon-valleys-fail-fast-mantra-is-just-hype.

4. We have borrowed this term from space exploration as a way to describe a happy medium. For a discussion of the earlier use of the term, see "The Goldilocks Zone," *Science@NASA*, October 2, 2003, https://science.nasa.gov/science -news/science-at-nasa/2003/02oct_goldilocks/.

5. Paul Michelman, "Do You Diagnose What Goes Right?" *MIT SMR* 58, no. 3 (Spring 2017), https://sloanreview.mit.edu/article/do-you-diagnose-what-goes -right/.

6. Bonnie McGeer, "Capital One Shortens the Machine-Learning Curve," *American Banker*, April 26, 2017, https://www.americanbanker.com/opinion/capital -one-shortens-the-machine-learning-curve.

7. Ibid.

8. Nicole Hemsoth, "Capital One Machine Learning Lead on Lessons at Scale," *Next Platform*, March 27, 2018, https://www.nextplatform.com/2018/03/27/cap ital-one-machine-learning-lead-on-lessons-at-scale/.

9. Bruce Dehning, Vernon J. Richardson, and Robert W. Zmud, "The Value Relevance of Announcements of Transformational Information Technology Investments," *MIS Quarterly* 27, no. 4 (December 2003): 637–656.

10. James G. March, "Exploration and Exploitation in Organizational Learning," *Organization Science* 2, no. 1 (1991): 71–87.

11. O'Reilly and Tushman, "Ambidextrous Organization," 74.

12. Fredrik Svahn, Lars Mathiassen, Rikard Lindgren, and Gerald C. Kane, "Mastering the Digital Innovation Challenge," *MIT SMR* 58, no. 3 (Spring 2017), https://sloanreview.mit.edu/article/mastering-the-digital-innovation-challenge/.

13. Peter F. Drucker, "The Discipline of Innovation," *Harvard Business Review*, August 2002, https://hbr.org/2002/08/the-discipline-of-innovation, accessed December 29, 2017.

Conclusion

1. Ilan Shrira, "Why 'The Wizard of Oz' Is the Most Popular Film of All Time," *Psychology Today*, June 4, 2010, https://www.psychologytoday.com/blog/the -narcissus-in-all-us/201006/why-the-wizard-oz-is-the-most-popular-film-all -time.

2. You can see a video clip of the ending at "There's No Place Like Home," from *The Wizard of Oz*, 1939, YouTube, video, 0:32, uploaded November 28, 2006, by manyhappyrepeats, https://www.youtube.com/watch?v=zJ6VT7ciR1o.

3. Several individuals have expressed this sentiment, including former secretary of state Condoleezza Rice. See, for example, Quin Hillyer, "Condi Rice, on Target in Mobile," *American Spectator*, November 14, 2011, https://spectator .org/36588_condi-rice-target-mobile/.

4. "The 50 Most Popular MOOCs of All Time," Online Course Report, accessed January 2, 2018, https://www.onlinecoursereport.com/the-50-most-popular-moocs -of-all-time/.

5. "How Companies Can Profit from a 'Growth Mindset.'"

6. Carol Dweck, "What Having a 'Growth Mindset' Actually Means," *Harvard Business Review*, January 13, 2016, https://hbr.org/2016/01/what-having-a-growth-mindset-actually-means.

7. Gerald C. Kane and Maryam Alavi, "Information Technology and Organizational Learning: An Investigation of Exploitation and Exploration Processes," *Organization Science* 18, no. 5 (September–October 2007): 786–812.

Index

Page numbers in italic refer to figures.